Detecting and Reducing Supply Chain Fraud

"No Shenanigans!"

Detecting and Reducing Supply Chain Fraud

NORMAN A. KATZ

LONDON AND NEW YORK

First published 2012 by Gower Publishing

Published 2016 by Routledge
2 Park Square, Milton Park, Abingdon, Oxfordshire OX14 4RN
711 Third Avenue, New York, NY 10017, USA

First issued in paperback 2016

Routledge is an imprint of the Taylor & Francis Group, an informa business

British Library Cataloguing in Publication Data
Katz, Norman A.
Detecting and reducing supply chain fraud.
1. Business logistics. 2. Fraud--Prevention. 3. Fraud investigation.
I. Title
658.7′2-dc23

Library of Congress Cataloging-in-Publication Data
Katz, Norman A.
Detecting and reducing supply chain fraud / by Norman A. Katz.
p. cm.
Includes bibliographical references and index.
ISBN 978-1-4094-0732-4 (hbk)
1. Business logistics. 2. Fraud--Prevention. 3.
Fraud investigation. I. Title.
HD38.5.K38 2011
658.4′73--dc23

2011052224

ISBN 13: 978-1-138-27006-0 (pbk)
ISBN 13: 978-1-4094-0732-4 (hbk)

Contents

List of Figures

List of Tables

Preface

Whew! I did it! My first book! If I was ever curious as to a good example of "a labor of love," writing a book has put my curiosity to rest.

How I came about creating my supply chain fraud detection and reduction business model stems from connecting several proverbial disconnected dots. My backstory will share a bit of my history but will also help you begin to picture and piece together the various concepts that I have woven together.

In around 1978, while in high school (also known as secondary school) at age 16, I was introduced to computer programming. That I could direct a machine on how to react to human interaction was fascinating to me. I took several computer programming courses in high school and I knew this field of study would be my major in university. At the time my alma mater, the University of Florida, offered computer information sciences curriculums through the colleges of engineering, business, and arts and sciences. I didn't care much for how computers were built or how their operating systems worked; I think I knew I wanted to use them as a tool to solve business problems.

After graduating from the University of Florida in 1985, I had the opportunity to work for several well-known US consumer brand companies, starting out as a programmer and working my way through the career path of programmer analyst, business systems analyst, and manager of information technology. Through my career I had the opportunity to be exposed to various Enterprise Resource Planning (ERP) systems and niche technologies such as barcode labeling and scanning and Electronic Data Interchange (EDI). My programming skills complemented my business operations knowledge and analytical thinking, enabling me in my problem-solving approach. After getting to the root cause of a problem, I could strategize the solution and implement the corrective actions right down to the data level.

In 1995, after suffering two back-to-back company downsizings and now facing my third (none of which were in any way my fault), I realized that I had to take my career – and destiny – into my own hands if I was ever going to have some reasonable measure of consistency in employment. I was at lunch one day with my barcode hardware vendor, who was telling me about some of her troubles. She could sell barcode hardware with no problems, but her company did not provide any software support, meaning that her customers were left on their own to figure out how to program the scanners, barcode printers for labels, and the integration into the existing ERP system or the stand-alone application software solution. While her company did not like her using outside resources, she had no choice because, she exclaimed, hardware without software was pretty useless. She used an outside group of software consultants to the displeasure of her company, but as she was the company's top sales representative, there was very little her employer could say about her repeated success year after year. She was very complimentary with regard to my knowledge of barcode scanner hardware and application software, and then stated her relative displeasure with the software resources she was using, finishing off her thoughts by stating to me: "I wish I had a guy like you I could use to help me."

Seizing the moment, I informed my lunch partner of my impending lack-of-employment situation. We agreed that we could work together: we had had a great customer–vendor relationship for several years. Within two weeks of that fateful lunch, I resigned from my employment at the time (I was not going to give the company the satisfaction of terminating my employment) at the end of November 1995.

For two years before this point, I had often thought that I could be in business for myself, but I didn't know what my special skill set was that would make me different from everyone else, a specialist versus a commodity. It was that fortuitous lunch meeting that was the beginning of the answer: barcode scanning applications. I labored over a company name and received lots of input from lots of people, but nothing stuck. After meeting with my accountant to discuss the idea of self-employment, I asked her what she thought regarding a company name, providing her with a few of the names suggested to me and what my own thoughts were.

I will never forget sitting in front of her desk as she leaned back in her large, well-padded, brown leather chair, staring straight up at the ceiling in silence for at least a full minute in deep contemplation. She eventually lowered the chair slowly and – looking me straight in the eyes – uttered the word "Katzscan."

I literally froze as it took several seconds to register, and then I broke out laughing. On January 1, 1996 my company Katzscan Inc. was officially born.

At first I started out merely specializing in barcode scanning applications such as inventory control, fixed asset management, and compliance labeling, partnering with my former barcode hardware vendor on the projects she presented. Using my programming skills, I wrote applications for the barcode scanners and PC-platform business applications as a compliment when my clients lacked the business software backend or an interface to their ERP system. I soon branched out to other specialty skills that I realized I possessed: EDI and my knowledge of supply chain vendor compliance. Again, my programming skills came in handy for stand-alone data mapping and data auditing applications as well as integration to my clients' ERP systems. True to her word, my former barcode hardware vendor and now business collaborator introduced me to several companies that each became great clients with repeat business over the following years.

As I was leaving my former employer in November 1995, a colleague at the time suggested I offer free software (Microsoft Word and Microsoft Excel) training classes at a local unemployment benefits office, with the reasoning that many unemployed professionals at that time did not know how to use spreadsheet or word processing software. As such, they would greatly appreciate adding these skills and then, when they found employment, they would be great business contacts and could help me to obtain consulting assignments. I proposed this idea to my nearest unemployment benefits office in Fort Lauderdale, Florida, and provided free weekly training for six months in Microsoft Word and Excel. (As I secured more business, my schedule became more hectic and I had to cease the training classes to accommodate my clients.) Sure enough, just as my former colleague had predicted, I did secure a great client from someone who sat in my classes and recommended me to their new employer, a jewelry manufacturer. The attendance started small (approximately 10–15 people), but as word spread I had sometimes 30 or 40 people – some happy to stand just to learn – in the little training room. During this time I met literally hundreds of people from all manner of professional backgrounds and I handed out business cards to the vast majority of them.

It was approximately one year after I had stopped my free training courses when I received a phone call from someone who had attended my classes and retained my business card: he had just opened his own private investigations company – having performed investigations when he served in the military –

and wanted to chat with me. We met at his office a few days later. He explained to me that the field of private investigations was becoming less "gum shoe" and more technical as the Internet was advancing forward as an investigations tool. I proposed that I become a private investigator (PI) under his agency as it would likely be better accepted by his clients, and he agreed; thus, my two-year apprenticeship to become a fully-fledged PI began. (Each state in the US sets its own requirements for PI certification; the license is regulated at the state, not the federal, level. Florida has a more stringent requirement to be a licensed PI than many other states in the US, hence the required two-year apprenticeship.) I learned a lot in our ten-year collaboration as I provided assistance and insights on a variety of business and personal investigation cases. I learned a lot not only about how private investigation cases are carried out but also about people in general.

After the collapse of Enron and World Com, my PI friend suggested that we get certifications specializing in white collar fraud, a direction he desired for this investigations company and that was well in keeping with my software and business operations skills. While his company did handle business investigations, they were usually tied to sometimes seedy and sometimes dangerous (e.g. cheating spouses, insurance deceptions) personal investigations, and he wanted to broaden his business model, moving away from the personal aspect of the investigations.

When people discover that I hold a PI's license, their first reaction is generally one of fascination and most comment that they have always wanted to be a PI. When I ask them why, their response is that it always looks cool on television shows, and they think that the fictional television shows must somehow be mimicking real life. This could not be further from the truth.

Paraphrasing my former PI colleague's response to those who inquired about his agency's services and were surprised by the amount of time and money necessary to invest in a professional investigation: Real-life private investigation is not a television show – crimes don't get solved in one hour.

There is nothing exquisite, posh, desirable, comfortable, or safe about spending hours and hours sitting in a parked car at 3:00 am in 90°F weather when it is also 90 percent humidity outside in the middle of a street in the middle of a high-crime neighborhood watching a house to catch a person cheating on their partner. Driving away to use the facilities of a local gasoline station or eatery could mean missing the opportunity to catch a person on camera perpetrating their act of infidelity.

It is very difficult to plan events and social engagements when at any time – if the private investigations agency is a busy one – the phone call could come in requiring an investigator to be at an airport waiting for someone to arrive or leave, catching them in the act, and at a specific location. Quite simply, as a PI your schedule is rarely your own and you are often required to be in unsafe environments where your very safety could be threatened by the persons you are investigating, their associates, or neighbors in reaction to a suspicious vehicle and person in their neighborhood.

Regardless of the case – cheating spouse, insurance fraud, or white collar crime – the suspects and their conspirators and collaborators may turn violent against the PI because their ability to carry out the fraudulent or illicit activities is being threatened by being exposed and therefore possibly terminated due to the presence of the PI.

PIs have no police authority – a common misconception about the trade. This is especially true of the responsibility of carrying a firearm: a PI's ability to defend himself or herself with a firearm is not the same as that of a law enforcement officer. Any evidence collected by a PI must be done via legitimate means and in full compliance with the laws of the land. A PI must always consider that the evidence collected may be presented in a court case; to collect evidence other than through legal and legitimate means could result in the testimony of the PI being dismissed and the evidence – no matter how supportive it is – thrown out.

After considerable searching on the Internet, we found the Association of Certified Fraud Examiners (www.acfe.com) and, after researching their credentials and those of other organizations and associations, we agreed that the Certified Fraud Examiner (CFE) credential was the one to get in order to help open the door to white collar fraud investigations.

The CFE study course is rather daunting: my version was a total of 2,200 pages which divided the four topics (accounting and finance, law, investigations, and criminology and ethics) into three manuals. I read the manuals twice because I didn't just want to understand the material; I wanted to get myself into the mindset of a fraud examiner. In this way I believed I had a better chance of passing the 2,000-question practice exam and 500-question certification exam. As I read and studied the material, I realized that – at the core – there was a tremendous focus on investigating fraud from an accounting point of view by examining the financial statements in order to find anomalies. To better understand the material, I began thinking of the frauds and investigations in terms of my clients' various businesses and their supply chain operations: since I was not an accounting or finance professional, I thought that if I related what I was reading to situations that I understood, I would grasp the concepts better.

As I continued through my second reading of the manuals with the supply chain perspective now firmly planted in my mind, I realized that trying to detect fraud by looking at the financial statements was more reactive than proactive. Good supply chain practices should audit (e.g. quality check) along the path, spotting something bad at the point of failure rather than letting the failure continue to infect the rest of the supply chain, potentially causing more damage as it flows through to the end.

Suddenly the term "supply chain fraud" popped into my head, along with the realization that I might have just conceptualized a unique viewpoint in detecting and reducing fraud. I thought that certainly the domain name supplychainfraud.com would have been taken and so I tried to put it out of my mind, but the thought just kept nagging me to the point of being a distraction! I left my study table and went to my computer to look up the domain name, which – to my surprise – was available in all extensions. I purchased the domain name on the spot.

As I continued my studies, I jotted down notes as I formulated my supply chain fraud business model. I achieved my CFE designation in November 2006 and began focusing on defining my supply chain fraud business model. I returned to what I had learned in CFE studies and began researching good governance and Sarbanes-Oxley Act of 2002 in greater detail, which led me to the Certification in Corporate Governance that I attained from Tulane University Law School. Now my supply chain fraud business model had a foundation upon which it could be built: good governance. I continued achieving my anti-fraud credentials when I was accepted for membership by the Association of Certified Fraud Specialists in 2008.

I was already an avid networker for Katzscan's core consulting services and promoted myself through speaking engagements and authoring articles in various trade publications. I pursued the same visibility opportunities with my new business model. I presented to various audiences that were composed of fraud examiners, accountants, internal auditors, PIs, and supply chain professionals. I wrote articles for both fraud and supply chain magazines.

As I write this preface in 2012, I have come to realize that my primary audience is not the supply chain professional as I thought it would be; rather, it is the internal auditor – someone with an accounting or finance background in charge of internal controls and the investigation of business transaction anomalies in their organization. As the time has passed since launching this business model, I have been asked to present to a variety of internal audit and accountant groups on a much more frequent basis than to a supply chain audience. More often than not, it is my experience that these audit and accounting professionals often do not have a full understanding of the supply chain, or even what a supply chain is, yet they all work in organizations or provide services to clients with typically large, complex supply chains. How then can one perform effective internal audits and create effective internal controls without an understanding of the organization from a supply chain perspective as it relates to both operations and technologies? The answer that my audiences have come to understand is that they cannot: the accounting viewpoint is not by itself sufficient for the detection and reduction of fraudulent activities in a supply chain.

This business model is a multi-faceted mix of technology and methodology, of supply chain practices and internal audit procedures, of legalities, behavioral activities, and ethicalities that, taken together, will enable an organization to more effectively detect and reduce fraud in its supply chains. This book is written for the supply chain professional and the auditor (internal and external), for the legal advisor and the ethical guide, for the accountant and the technologist. I have used examples that reflect bad supply chain behavior and lapses in internal controls, that highlight questionable ethical behavior and examine interpretations of law, and that describe new uses for old technologies and mix operational improvement with fraud detection.

Detection and reduction share a common characteristic: the dedication to both starts at the top of the organization and is embraced all the way down. Neither can be started – much less exist – without senior executive management being behind both initiatives, which includes leading by example. People are the cause of the problems and only people can be the champions of the solutions. Doing what is right is probably the hardest thing that anyone can be asked to do by someone else or tasked to do themselves, but doing the right thing must be done.

The detection and reduction of supply chain fraud – especially for national or global supply chains – is a multi-dimensional problem that requires expertise in multiple disciplines in order to be resolved. It is not just about one viewpoint to a problem, but multiple perspectives in parallel.

Reviews for *Detecting and Reducing Supply Chain Fraud*

Auditors seek to confirm that transactions are authorized, accurate and complete. By definition all fraud is non-authorized. Norman Katz shows, via multiple stories and examples, those areas where authorization controls fail – from their design – and how to create robust authentication processes that include prevention and detection of control failure. A must read for Board members.

Frederick D. Cox, Vice President Information Security and Privacy, Seacoast National Bank and CEO, FDC Associates, LLC

Norman Katz's "eye opening" examination of fraud is required reading for anyone involved with supply chain operations. Easy to read and chock full of practical advice, the fraud detection and reduction methods will increase the efficiency and effectiveness of supply chain governance, risk management and compliance processes. This is a significant contribution to the field of operational excellence!

Doug Ross, President and founder of Principle Dynamics

Introduction

Before we can go about trying to detect something, let alone look at reducing how often it occurs or its severity should it occur, we need to understand what it is we are looking for and where it is likely to be found. We need to identify its characteristics so as to determine how it might appear when discovered. An examination of its traits may help us devise a means of forcing it to reveal itself under specific (and in all likelihood somewhat controlled) conditions. An analysis of its habits can help us ascertain favorable environmental factors where it might be located and even thrive. We are the hunters. Can our prey camouflage itself, thus rendering it difficult to be seen? Can it disguise itself to look like something else? How else can our prey elude capture?

Whether we have donned deerstalkers and cloaks or white laboratory coats, whether we have spyglasses in hand or our hands on microscopes, we are investigators on the trail of disruptors and their counter-productive activities. We will need to collect evidence and perform detailed analysis of data. We will need to interview possible suspects, innocent bystanders, witnesses, and unfortunate victims. We will trace leads to determine whether they take us down dead-end paths or are truthful revelations. We will pit our wits against those who seek to outsmart and deceive us. We will shift our mindsets away from what we know is right and true in order to place ourselves in the state of mind of the perpetrators. We will call in experts to help us understand the intricacies of specialized disciplines as we piece together the puzzle parts to formulate our conclusion and propose a solution.

The detection and reduction of supply chain fraud requires an understanding of what the (or a) supply chain is, what constitutes fraudulent behavior, and where fraudulent activities can be located. We can then look at reduction technologies and methodologies as well as factors that enable fraud to exist and foster (though fester may be a more appropriate term).

As you read this book, please keep in mind two very important points:

- This book is not a how-to guide for perpetrating fraud or any other disruptive behavior.
- I am not a lawyer or accounting professional and cannot be relied upon for legal or financial advice.

It is a "whodunit" that can span the globe and involve players from all walks of life and at all levels of career hierarchies. It is a race to prevent bad behavior from taking control. It is a chase to mitigate the damage done when a disruptive action infects an otherwise healthy system. It is a challenge that pits the intellect of defenders of good against the scheming minds of ne'er-do-wells.

This is your guide to the understanding, detection, and reduction of supply chain fraud, and the game is afoot!

PART I

Understanding Supply Chain Fraud

1

What is the Supply Chain?

Without purposefully adding yet another definition of the supply chain to the existing ones, it is important in the discussion of supply chain fraud to broaden the scope of what the supply chain represents. As the ideology and applicability of the supply chain concept continues to grow, the definition needs to encompass the overall spectrum of business activities without being tied to a particular industry type. My definition of the supply chain is as follows: *The movement of something between a supplier and a customer from start to finish.*

Now let us analyze the components of this definition. First, something moves. It does not really matter what it is: raw materials, components, finished goods, monies, services, data or information (the latter being the more intelligent version of the former), and documents. Conceptually, the supply chain is a line of interconnected links and it is along this path that something moves from one link to the next in sequential order.

Second, something moves between a supplier and a customer. Traditionally the concept of the supplier and the customer were of entities outside the walls – physically even more so than literally – of an organization. Suppliers provided something that an organization purchased and customers purchased something that the organization provided. But in a broader sense the supplier and customer could be internal or external and could be entities belonging to the organization, independent entities not belonging to the organization (whether located internally or externally), or a combination of these. The existence of the internal supplier and the internal customer relationship – both belonging to the organization – has been around for a long time. Imagine an organization's Information Technology (IT) and Human Resources (HR) departments. These departments supply goods and services to the other departments in the same organization, which represent their customers. (The IT and HR departments are examples of entities who are both suppliers and customers as it is likely that each of these departments relies on suppliers for professional services

Table 1.1 Internal Supplier–Customer Relationships

Internal Supplier	Internal Customer
Receiving	Quality Assurance
Quality Assurance	Inventory Control
Inventory Control	Manufacturing
Purchasing	Accounting (Payables)
Sales Order Processing	Accounting (Receivables)

and goods. Most suppliers are at some point customers themselves and most customers are suppliers of a product or service.) Similarly, there are other internal supplier–customer relationships, such as those identified in Table 1.1, that exist in a manufacturing organization.

In industries such as the automotive and electronics industries, some independent suppliers will set up shop within their customers' manufacturing facilities, representing an example of an external supplier being located within the customer's physical walls. This closer supply chain relationship can represent greater control over inventory levels and can reduce replenishment cycles, resulting in significant efficiencies for both trading partners.

Third, something is moved between a supplier and a customer from start to finish. How something is started and finished is rather subjective and is dependent upon which link we are located at in any given supply chain. During a manufacturing process we may look at a particular work-in-progress assembly or fabrication step, or we may consider the testing of samples by a quality assurance department prior to the acceptance of raw materials or finished goods. In a financial institution this could represent the movement of a loan contract from one department to the other to determine the requestor's viability for the loan amount, or it could represent the transfer of money when the bank provides loan funds to a customer.

It is important to note that:

a) there is a handoff of something from the supplier to the customer;

b) there is a transfer of the responsibility for that something from the supplier to the customer; and

c) the move is complete or whole, in that there should be no expectation that the move is not permanent in nature for whatever reason.

It should be unacceptable to allow the move of something that was not of appropriate quality, especially if there is an expectation that the movement of something of less than appropriate quality could result in its return to the supplier or subsequent disruption of the supply chain. The determination of what constitutes something of appropriate quality depends on the particular industry supply chain and the requirements of the given customer, e.g. manufacturing versus financial. In manufacturing, raw materials must pass specific tests for purity and tolerance to engineering specifications in order to be deemed first-quality. In a financial institution, a document that is not first-quality may be missing necessary information or approval signatures that should exist at a particular supply chain link and are required to move the document on to the next step.

The gaps in responsibility and completeness are root causes of finger-pointing when problems arise. Each party – the supplier and the customer – can look to blame the other for the condition of the something that was moved that resulted in a particular failure, or the need for redundant efforts, such as the customer's return of something to the supplier. It is in this gap that organizations suffer excessive costs due to higher than necessary overheads from the results of failure and inefficiency, e.g. if the customer became burdened with the responsibility of fixing problems that the supplier should have caught and corrected before the move. It is also within this gap that fraud can occur, just as it can occur within a link itself. Reducing the amount of movement in a supply chain can equate to reducing the number of incidences of fraudulent activity.

Enterprise Resource Planning Systems

Originally, Materials Requirements Planning (MRP) software applications helped manufacturing companies determine how much of particular raw materials were needed, based on planned or projected inventory levels of finished goods against the organization's overall capacity to make something. Bills of materials, operations, and labor represent the key aspects of manufacturing finished products: The bill of materials lists which ingredients are needed, the bill of operations describes the order of the manufacturing steps, and the bill of labor retains the amount of human effort needed per operation.

As businesses became savvy, the need to interface MRP systems with the software applications of the other areas of the business, such as the financial areas, became more necessary. Traditionally, companies would run "best-of-breed" software applications to support the distinct functional departments. (In my view, the "best" software or solution in general is often the most functionality that one can afford.) The "best" MRP software ran the manufacturing side of the business, the "best" accounting software handled all the financial needs, and the "best" order entry software ran the sales order processing department. These disparate systems would pass batches of data files between each other, typically at night, to provide the necessary updates. As the real-time needs of businesses grew, so did the realization that it was necessary for one comprehensive software application to handle all the operating, administration, and accounting requirements of the business, linking together all transactions and without the need for batch data processing and the necessity of running an organization based on reports from different software applications. Thus, MRP (and later MRP II) software applications morphed into the modern Enterprise Resource Planning (ERP) systems.

ERP systems may be on the edge of an evolution themselves: New buzz-phrases such as "Beyond ERP" and "Enterprise Architecture" are emerging as the next generation of enterprise applications. If what goes around comes around, there seems to be a movement toward establishing "best-of-breed"-type enterprise applications to replace single ERP systems that are too weak in some functional areas needed by clients or that are demanding too much financial investment in terms of maintenance and license fees. Regardless of what the future holds for the enterprise application, there will still be a need at the core – whether it is one application or several interconnected applications – to retain foundational business information and track financial transactions. Thus, in the discussion of the supply chain fraud referencing, an organization's (traditional) ERP system changes nothing; business is still business and we still need specific software applications to hold our business information and capture our business transactions. Nor does it necessarily matter whether the core applications are run on mainframes, client-server architecture, or in a cloud.

These manufacturing concepts are applicable to other sectors such as the legal, finance/banking, education, and health care sectors. For example, the completion of a financial transaction requires materials (e.g. documentation), various operations, and labor at each operational step. Note that my definition of the supply chain is not targeted at any particular industry but is applicable generically. Supply chain fraud is not an industry-specific problem. Readers should be drawing parallels between traditional manufacturing concepts and their own industries because these parallels do exist. Examples in this book refer to manufacturing organizations because of the extra level of detail that can be showcased and the fact that manufacturing cuts across so many industry types, but there is equal applicability to distributors who purchase finished goods, banking and investment firms, law firms, hospitals and other medical practices, educational institutions, etc. What a manufacturer or distributor calls its "customer" (the outside entity it sells its finished goods to) a law firm calls its "client," a hospital calls its "patient," and a college or university calls its "student." Fundamentally, each of these different industries runs its own form of an ERP system, and in fact some ERP software companies offer industry-specific solutions based on their core application product.

The supply chain operations of an organization are generally defined by the functions of its ERP system. Overall, ERP systems are the comprehensive software applications where organizations (a manufacturer is used in this example) store their business entities (see Table 1.2) and generate business transactions (see Table 1.3) that are recorded against their respective accounting ledgers.

An organization's standard operating procedures manual on how to use its ERP application in support of its business processes should represent the flow of its supply chain operations. The ERP system is a critical supporting technology of an organization and its supply chain operations, and is also a central technology to the detection and reduction of supply chain fraud.

Table 1.2 ERP System Information

ERP Entities
Customers
Suppliers
Raw Materials*
Finished Goods*
Components*

Table 1.3 ERP System Transactions

ERP Transactions
Purchase Orders
Sales Orders
Work Orders
Invoices
Pick Tickets
Accounts Payable
Accounts Receivable
General Ledger

Note: * Including quantity on hand, quantity by location, and value.

The Holistic Supply Chain

The traditional reach of the supply chain has to be redefined in the discussion of supply chain fraud. Suppliers are no longer local and thus are more difficult to connect with and exert control over. Outsourcing has become a commonplace strategy for the cost-effective acquisition of capabilities, especially those not within the core expertise of an organization. Even under the same organizational umbrella, it is no longer unusual to find divisions or departments housed in different facilities across a large geographic area. Employee telecommuting is gaining acceptance and can reduce the level of oversight (for this, read internal controls) that exist at the organization's physical offices. The supplier–customer relationship is a mix of internal and external entities and is sometimes a combination of both.

This holistic vision of the supply chain – a perspective that is not bound by an organization's walls (whether literal or figurative) – is of critical importance because fraud can happen anywhere, can by committed by anyone, and can be carried out in collusion with internal and external entities, regardless of whether they belong to or are independent of the organization. An investigative trail cannot be ignored just because it is either within or outside of the organization. The global economy does not allow for such a myopic view of supply chain fraud detection. Reducing fraud must be undertaken across the entire length of the supply chain, even if (or especially if) it stretches across the globe.

It is worth remembering that Ponzi schemes – such as the $65 billion fraud perpetrated by Bernie Madoff – have supply chains. The perpetrator must rely on a supply of investors. There are likely to be fraudulent documents (a website, investment reports, and financial statements, for example) to conjure up in support of the scheme. Office space and supplies, employees, computers, sales and marketing materials, and so on might have to be established to make the façade seem real. If the money is even partially invested, investment firms may be involved, whether as innocent bystanders, victims, or co-conspirators. The money has to be stored somewhere, so one or more banks may be involved as unaware participants, fraud targets, or involved collaborators. The same money may have to be moved around between different financial institutions to perpetuate the fraud and create a veil that masks what is really going on. The tracing of funds in such a scheme follows the money's supply chain.

So encompassing is the concept of the supply chain as an ideology that it is finding application in new and different areas of discipline. As reported in the November 9, 2009 issue of *Information Week* magazine in the article "Secure The Cyber Supply Chain" by J. Nicholas Hoover, researchers from various universities are collaborating with members of different companies and associations in the application of supply chain management principles to the security of computer systems and the software programs and data that reside within them. This also expands the definition of what we think a supplier or customer should be. Instead of merely limiting the definition of a supplier or customer to that of another organization, a department, or a person, a supplier or a customer in a data supply chain could also be a software application or hardware device. Software need not be complex ERP systems but could be much more focused applications, such as those used in programmable logic controllers. Hardware devices could include anything from search engine appliances to barcode scanners to robots or other types of automated equipment common in manufacturing environments.

Virtually every organization – from a mom-and-pop business to global conglomerates – has a supply chain of some type, regardless of whether it is short or long and without restriction to the industry or sector it is part of.

In the executive summary "Values-Based Food Supply Chains: Strategies for Agri-Food Enterprises-of-the-Middle," Steve Stevenson of the University of Wisconsin-Madison's Center for Integrated Agricultural Systems and Rich Pirog of Iowa State University define a "food supply chain" as follows:

Inputs → Producer → Processor → Distributor → Wholesaler → Retailer → Consumer

Food supply chains can have as inputs animal feed, plant fertilizer, veterinarian services, labor, and machinery to sustain and develop the animals and plants that will move forward through the food supply chain.

How Do Supply Chains Work?

This question is in all likelihood the topic of many a book and article, so I will keep it brief, but it is another foundational concept that is important when looking at supply chain fraud; thus, it warrants an overview. Essentially, supply chains work by the controlled and sequential firing of triggering mechanisms that individually cause an event to occur. (This is analogous to how gasoline vehicle engines operate. The spark plugs do not fire all at the same time, but rather in a particular sequence based on the engine type and manufacturer. This is controlled by either the distributor [more common in older vehicles] or via computer [more common in newer vehicles]. The timing and sequence that control the firing of the spark plugs is critical to the engine's operating performance.) Typically, one event cannot occur until the previous event has completed the sequential order. This is not to say that multiple events cannot be activated by the firing of a single trigger mechanism, as some supply chains are more complex than others, but for the purposes of this text I will restrict the supply chain example to one-on-one events. The firings should be controlled, in that there are rules that govern when a triggering mechanism will fire, e.g. a trigger mechanism may not fire until the previous trigger mechanism has fired. Trigger mechanisms can also be activated when situations arise that are outside the normal rules of business procedure. (Management by exception is much better than exception management, so hopefully these outside-the-norm types of trigger mechanisms are rarely activated.) In terms of our supplier–customer relationship, there must be some trigger event that causes a supplier to move something to a customer. Whether the movement is due to a "pull" from the customer or a "push" by the supplier is a matter of debate, and one that is somewhat beyond the scope of a discussion on the subject of fraud. The fact is that regardless of whether a relationship is defined as a push or a pull – and there are valid business situations for each – a trigger mechanism must fire to cause the movement of something from the supplier to the customer as defined at a given link. Each link of the supply chain is activated by some type of trigger event. The misfiring of a trigger event can cause disruption to the supply chain and may be due to the result of fraudulent activity. As such, an examination of supply chain performance on the whole and at each step adds visibility in detecting supply chain fraud.

A trigger mechanism can be sensory (verbal, visual, or tactile) or systematic. A verbal trigger mechanism can be as simple as two people speaking – or yelling – at each other (hopefully not at the same time, but rather in the form of a conversation). Two workers on an assembly line might communicate to each other as a signal of the handoff and receipt during a manual work-in-progress step. Visual trigger mechanisms are sometimes used by lean practitioners to signal the need for replenishment. An empty inventory shelf could be a clear sign that more goods need to be manufactured to fill the space. The raising of a red flag or the manual flipping on of a light at a manufacturing work-in-progress operational step could be a signal that more raw materials are needed to continue without interruption.

Stamping a document with a check mark or OK symbol (indicating that the document is correct) versus a document with an X (indicating that the document is in error) are good signs of the document's status and are visual indicators that inform as to which path the document will take along its supply chain. Systematic trigger mechanisms include automated email notifications and red warning or "adverse situation" highlights on software dashboards. (Though the latter is visual, it is systematically generated – via an automated process – as opposed to one that is manually initiated.) If an organization's accounts receivables balance goes beyond a certain threshold, there should be an alert to the appropriate accounting manager via a daily report, email, or dashboard highlight. Actions to be taken can include ensuring the collection staff are taking the appropriate steps to contact past due customers. A review of the organization's credit policies may reveal that terms are being provided to customers with credit histories that reflect too much risk, and this will necessitate the organization revising – raising the bar – on what it will minimally accept with regard to a customer's credit report.

It is important to ensure that the trigger mechanism is appropriate for the need. Portable handheld barcode scanners can be programmed to emit selected beeps at different volumes and pitches to provide alerts to the user. To alleviate the need to look constantly at the handheld display – to determine a good barcode scan versus a bad barcode scan or the results of data validation – the beeps provide an audio feedback. There is much greater efficiency and ergonomic comfort in the user listening for a particular beep than constantly having to make physical movements in order to inspect the display visually. This audible trigger mechanism works well, except in noisy environments where the beeps cannot be heard. This is why in around the mid-1990s (from what I recall) handheld barcode scanner manufacturers included vibration feedback in their devices. (Vibration feedback is now common in cellular telephones, especially those with touch-screen displays.) A dashboard notification or printed report might be insufficient for someone who is often not at his or her desk or who travels frequently.

Inbound and Outbound Supply Chains

Without the impediment of walls – physical or virtual – to impair our vision and cloud our judgment, we can simply look at supply chains generally as being inbound or outbound in direction. For a manufacturing or distribution company, an inbound supply chain could resemble Figure 1.1 and an outbound

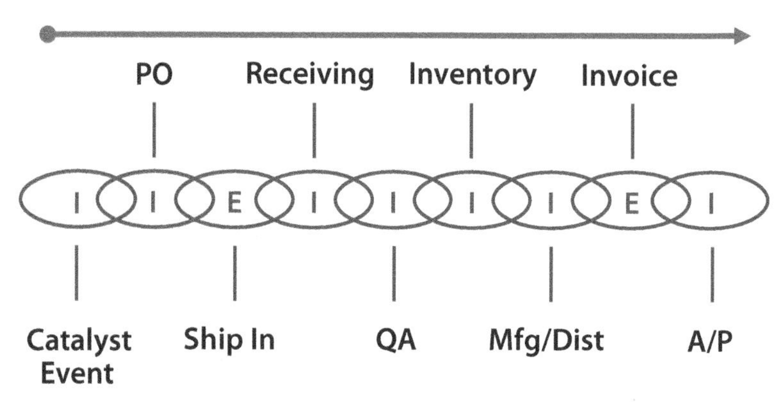

Figure 1.1 Inbound Supply Chain Example

supply chain could resemble Figure 1.2. (An examination of each link in the inbound and outbound supply chains and the frauds that can occur there are detailed later in the book.)

With regard to the inbound supply chain example:

- A trigger mechanism is fired that causes the inbound supply chain to come alive and become active. The event that occurred is most likely the need to acquire something, such as raw materials, components, or finished goods.

- A purchase order to a supplier is generated to authorize the acquisition of what is needed.

- A supplier ships based on what was on the purchase order.

- The inbound shipment is received by the customer.

- A quality assurance check is performed on either part of or all of what was received.

- What was received and passes the quality check goes into inventory; what does not pass the quality check goes into a quarantine area.

- From inventory, the raw materials would normally go to a manufacturing process and then the finished goods sent to inventory or directly to distribution. If what was purchased did not

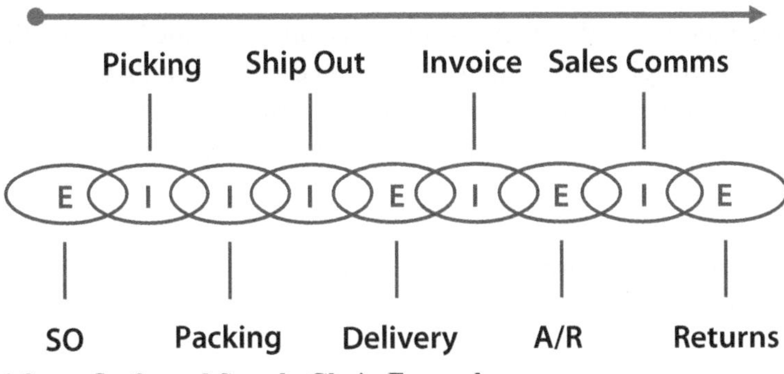

Figure 1.2 Outbound Supply Chain Example

require any manufacturing, it would go direct to distribution from inventory.

- An invoice is received from the supplier.
- The supplier's invoice generates an entry in accounts payable.

With regard to our outbound supply chain example:

- A sales order is received, which is the authorization to ship something to a customer.
- The contents of the sales order are picked, packed, and shipped out. (This is indicative of a build-to-stock situation; if the organization was build-to-order, there would be a preceding manufacturing process.)
- The customer receives delivery of what was shipped.
- The customer is sent an invoice from the supplier.
- An accounts receivable entry is created against the invoice, which relies on the receipt of an external payment in order to be closed.
- If the organization employs sales representatives, it may be responsible for commission against the customer order.
- The organization may be required to deal with customer returns if any part of the order was unsatisfactory or in error for some reason.

In both the inbound and outbound supply chains, it should be noted that more of the activities originate internally (as indicated by "I") within the organization compared with those that are externally (as indicated by "E") originated by unrelated or independent entities. This holistic supply chain perspective is significant because:

a) more of an organization's supply chain is under its control than not;

b) the organization has more opportunities for being disrupted due to internal fraudulent activities; and

c) it means that any fraud investigation must look across the entire spectrum of an organization's supply chain activities with a particular focus on what is happening inside.

The inbound and outbound examples may differ, depending on the complexity and specific industry of the organization. For example, delivery of a customer order might rely on a third-party logistics provider, the use of independent parcel carriers or trucking companies, or the use of an organization's own transportation fleet. Regardless of the means of delivery or the level of organization ownership, something is handed off from a supplier to a customer and this particular transaction happens outside the organization's walls. In this example the delivery is made to an external customer (who may or may not be independent of the organization) and is applicable whether we are profiling a consumer's online purchase of a product or the delivery of fuel to a gasoline station.

There are challenges to be faced when performing internal versus external fraud investigations. External investigations may be hampered by the inability to access certain information from suppliers and customers. Revealing that a fraud investigation is being performed may not be information that an organization wants its suppliers and customers to know. Internal investigations may be blocked due to politics at various levels of management. Cooperation from suppliers and customers in a fraud investigation might be more forthcoming, especially if the supplier or customer is a potential victim. On the other hand, if the customer or supplier is an active participant, it is likely to be resistant to cooperation. Internal cooperation might be difficult because someone's job (that of the fraudster and any co-conspirators) might be jeopardized.

2

What is Fraud?

OK – we all know that fraud is bad, but what is it really? It is likely that there are just as many definitions of fraud as there are of the supply chain, and again I had some misgivings about adding more to the mix. Fortunately this time I can make amends and use an existing definition. In this case I found a definition of fraud on *dictionary.com* that I think really encompasses what fraud is all about:

> *Deceit, trickery, sharp practice, or breach of confidence, perpetrated for profit or to gain some unfair or dishonest advantage.*

What I believe makes this definition so especially appropriate to the discussion of supply chain fraud is the recognition that fraud is a breach of confidence. (I think that the definition could leave out "deceit," "trickery," and "sharp practice" as they all represent some of the means to the end in terms of how confidence can be breached.) Persons inside and outside the organization can breach confidence in many different ways, as this book will highlight. The gain of a fraud can be direct or indirect. An example of a direct gain is the theft of goods or monies by an individual. An indirect gain can be realized by an individual who uses bribes to secure favorable votes for his or her employer during a public works bid or proposal process, and in this latter case the person can gain an unfair advantage over the superior bids of competitors through the perpetration of fraud.

Not as obvious in this definition but very important nonetheless is that fraud is carried out with intent. Fraudsters intend to breach confidence in order to achieve gain. Accidents do not happen on purpose; a person does not accidentally perpetrate fraud. If fraud is carried out in order to fix a problem, the gain may be the relief of the burden. In such a case the gain is likely to be more emotional than it is material: Even if money is gained as a result of the fraud, the likely use of the money is to relieve an emotional burden of want or need of something. (The relief may be temporary or the burden may shift from dealing with the original problem to dealing with the guilt, cover-up, or other repercussions of the fraud. Fraudsters do not always think about the long-term effects of their actions because their perspectives are usually more short-term, due to greed or the need to rectify a situation that has spiraled out of control.)

By the same token, willful negligence does not excuse a person from fraudulent behavior because the negligence was *willful* (deliberate) and not accidental. This raises an interesting question: Similar to asking if a person can be set up to fail, can a person be set up to unknowingly perpetrate fraud? Yes. If an organization willfully does not equip or train its employees properly or directs or coerces an employee – through the employee's own lack of knowledge or capabilities or under pressure of risk of job loss – to commit an act of fraud, was it not really the organization that failed and should be held accountable rather than the employee? This question will be examined in more detail later in the book.

The US Internal Revenue Service (IRS) is the federal government agency whose primary mission is to collect taxes. The IRS encounters both individuals and organizations who attempt to reduce – if not eliminate – their tax burden through the exploitation of loopholes and questionable interpretations of the tax law. When the evasion of federal tax responsibility becomes illegal, it does not necessarily a leap of faith to believe that fraudulent behavior is involved. The definition of corporate fraud by the IRS, found on its website (www.irs.gov), includes the following:

> *These schemes are characterized by their scope, complexity, and the magnitude of the negative economic consequences for communities, employees, lenders, investors, and financial markets.*

Some may smirk at this definition as being one that typifies the vagueness we have come to expect from the wordsmiths of our respective governments. The definitions of "scope," "complexity," and "magnitude of economic consequences" all seem quite subjective. After all, one person's pittance is another person's wealth. But let us not dwell too much on how the IRS quantifies corporate fraud; rather, let us look at who the potential victims are: communities, employees, lenders, investors, and financial markets. I submit that the term *communities* include all consumers of an organization's products or services, the external customers and suppliers, and the cities or towns where the organization has established itself. Fraudulent behavior has been proven to have caused injury or death to the consumers of an organization's products. Fraudulent behavior that results in the closure of an organization's headquarters, branch offices, manufacturing facilities, suppliers, or customers can have a devastating negative financial impact on a local economy. Where in an organization would such frauds occur that could negatively impact all or part of such a diverse audience? The answer is the supply chain operations, which include both the business processes and the supporting technologies. Supply chain frauds can be perpetrated by white collar and blue collar employees alike in corporations (public companies or private enterprises) and government agencies. (The definition by the IRS excludes government agencies, in all likelihood because they do not pay taxes and thus are not within the purview of the IRS. The US General Accounting Office [GAO] is the independent agency that investigates the spending of taxpayer dollars by the US government.)

The UK Fraud Act 2006 states that fraud is defined as:

- False representation.
- Failing to disclose information.
- Abuse of one's position.

With regard to false representation, the Fraud Act informs that the deceit may be express or implied. Thus deceitfulness could be used to persuade an employee to carry out an illicit act that the employee would not normally perform under more honest and truthful circumstances.

Supply Chain Fraud Categories

Section 7 of the UK Fraud Act 2006 is titled "Making or supplying articles for use in frauds" and states that:

> *A person is guilty of an offence if he makes, adapts, supplies or offers to supply any article –*
>
> *(a) knowing that it is designed or adapted for use in the course of or in connection with fraud, or*
>
> *(b) intending it to be used to commit, or assist in the commission of, fraud.*

As noted in this section and as highlighted throughout the examples used in the book, fraudulent behavior includes making something used in the execution of fraud. This can include false documents, such as court orders, financial statements, websites, purchase orders, and invoices, as well as false products. As noted in s. 6 of the Fraud Act: "A person is guilty of an offence if he has in his possession or under his control any article for use in the course of or in connection with any fraud." Even if a person did not create a fraudulent article (an object or document), the act of possessing it can still get a person into trouble.

With the holistic perspective of the supply chain, we can categorize the different frauds that can occur. While not intended as being a complete list, supply chain fraud categories include the following.

ASSET MISAPPROPRIATION

1. Theft: Taking something that is not yours.

2. Embezzlement: Taking something that is not yours but you were entrusted to keep hold of. Robbers thieve, banks embezzle. The difference is that we never trusted the robbers with the care of our valuables in the first place; banks on the other hand are institutions that we trust to safely hold our money and other valuable assets.

3. Abuse: Asset abuse includes the deliberate failure to perform schedule maintenance that in all likelihood will reduce the life of the asset. Utilization of an asset in extreme conditions that are beyond its recommended limits will likely abuse the asset and reduce its usable life and depreciable value.

4. Misuse: Asset misuse includes the utilization of an asset for purposes other than those for which it was designed. This is one reason why neither shoes nor screwdrivers are viable substitutes for hammers and neither is a firearm – especially if it is loaded – though perhaps for more obvious reasons.

> During a software consulting assignment, I was asked to investigate manufacturing equipment failures because it was believed that the failures were computer control-related. The manufacturing machines on the (un-air-conditioned) shop floor were each controlled by command consoles that were essentially large circuit boards measuring approximately four feet square. The enclosed consoles

generated so much heat that large fans were designed into them to allow air to circulate and dissipate the heat, much like the fan on the rear panel of a personal computer. On the opposite console wall of the fan was an air filter that looked exactly like it came off my first automobile: A 1972 Pontiac LeMans.

Upon opening up the console covers, I found the master control boards covered in layers of dust. The fans were equally covered in dust and were not able to work effectively. I opened up one of the air filter housings and removed the air filter itself, which was covered in charcoal-black dirt. (It had obviously never been cleaned or replaced.) The maintenance supervisor informed me that he and his staff were performing regularly scheduled maintenance on the manufacturing machines, including the command consoles. Holding the air filter behind my back, I walked into the CEO's office and asked him if his maintenance staff were trained in the proper care of his six-figure manufacturing machines. The CEO swore that his staff were trained and were doing their job – after all, that was what they were getting paid for. I showed him the air filer and his jaw dropped, his mouth gaping wide open. I told him that, while I was not an engineer, I suspected that the machine manufacturer put fans and air filters in the consoles for a very good reason. I further informed him that his machine failures could be due to overheated consoles, which were causing some of the sensors to record information incorrectly. The CEO called his maintenance supervisor into his office for a private chat. The next day new air filters were in all the machine consoles, having been easily acquired – at my suggestion – from a local automobile parts store.

Quite simply, the machine consoles were taking considerable mistreatment due to overheating by maintenance personnel who were either deliberately not doing the job they were paid to do or were not trained on how to do their jobs properly. The deliberate failure to perform regularly scheduled maintenance is a critical issue for organizations that produce food and pharmaceutical products, because the failure to perform certain cleaning functions can result in the growth of bacteria, which then can be introduced into the finished good and cause harm to the consumer. By the same token, organizations have the responsibility to properly train people in how to do their job correctly and as defined by their job description, and then to ensure that the proper checks and balances are in place to ascertain whether the employees are performing their responsibilities.

CONTRACT AND PROCUREMENT

1. Bid rigging: This occurs when multiple bidding parties agree to fix prices or other contract terms, such as to make one party's bid a (clear) winner. The parties involved may agree to rotate as to who wins the next bid so as to spread the wealth between all parties. Kickbacks from the winning party to the other participants may be involved.

According to the US Department of Justice's "An Antitrust Primer for Law Enforcement Personnel" dated April 2005, price fixing is "any agreement among competitors which affects the ultimate price or terms of sale for a product or service." The parties do not have to agree to set the price exactly the same as each other, and price fixing can be directed toward setting prices either artificially high or low. In 1996, through the combined efforts of the Department of Justice and the FBI, Archer Daniels Midland (ADM) was fined separately $70 million and $30 million for its involvement with international conspirators in the lysine and citric acid price-fixing conspiracies.

2. Phantom bids: This covers the submission of fake bids from either real or *fictitious* companies used to make actual bids look more favorable. This taints the overall bidding process by effectively narrowing the choice of viable suppliers. The creation of phantom documents is applicable to financial fraud cases such as Ponzi schemes.

3. Nepotism: Giving favorable consideration to family members or close acquaintances.

4. Substitution: Providing the promised product but not according to first-quality specifications or substituting the promised product (without buyer approval) for something else not fully compatible with or comparable to what was promised. This could certainly be extended to frauds that offer something but deliver nothing.

A former official in a town in South Florida was caught and convicted of stealing $500,000. The fraudster had set up a dummy company that submitted bills for services to the town. As an official with the town, the fraudster had the authority to approve invoices and directed that he approve all invoices from the dummy company he had set up in secret. It was finally determined that while the town had been paying for the aforementioned services, it was seeing no results and an investigation exposed the reason why.

In another more egregious case that was reported in 2008, an individual in Miami Beach, Florida was indicted by the US government for acquiring Chinese-made ammunition in Albania and selling it to the US military, but as Albanian and not Chinese ammunition. Selling Chinese-made ammunition to the Pentagon is illegal based on an arms embargo enacted in 1989. Some of the Albanian ammunition was of such poor quality as a result of how it was stored that it failed to perform as required by military specifications. Imagine being a soldier in combat and not being able to protect yourself, protect your comrades, or defend innocent civilians due to having to use unstable ammunition that could misfire or cause weapon damage and thus cause injury.

FINANCIAL STATEMENTS

Financial statement frauds can include outright fictitious numbers on falsified balance sheets, income statements, profit and lost statements, and so on. These frauds include the deliberate exclusion or misrepresentation of information on the general ledger that is used as a foundation to create financial reports. Financial statement frauds can be used to secure loans and lines of credit at a level that the organization cannot legitimately apply for based on its financial reality. Though more relevant to business activities than personal matters, mortgage frauds could be considered a subset or offshoot of financial statement frauds because mortgages rely on the submission of our own personal financial statements.

In complaint CV-03-J-0615-S dated March 19, 2003 filed by the US Securities and Exchange Commission (www.sec.gov) in the US District Court, Northern District of Alabama, against HealthSouth Corporation (HRC) and Richard M. Scrushy, former Chief Executive Officer and Chairman of the Board, the introduction begins with the first two allegations:

1. *Since 1999, HealthSouth Corp. ("HRC"), one of the nation's largest healthcare providers, has overstated its earnings by at least $1.4 billion. This massive overstatement occurred because HRC's founder, Chief Executive Officer and Chairman of the Board, Richard M. Scrushy ("Scrushy"), insisted that HRC meet or exceed earnings expectations established by Wall Street analysts. When HRC's earnings fell short of such estimates, Scrushy directed HRC's accounting personnel to "fix it" by artificially inflating the company's earnings to match Wall Street expectations. To balance HRC's books, the false increases in earnings were matched by false increases in HRC's assets. By the third quarter of 2002, HRC's assets were overstated by at least $800 million, or approximately 10 percent of total assets. HRC's most recent reports filed with the Commission continue to reflect the fraudulent numbers.*

2. *Despite the fact that HRC's financial statements were materially misstated, on August 14, 2002, Scrushy certified under oath that HRC's 2001 Form 10-K contained "no untrue statement of a material fact." In truth, the financial statements filed with this report overstated HRC's earnings, identified on HRC's income statement as "Income Before Income Taxes And Minority Interests," by at least 4,700%.*

Based on hearing first-hand evidence from William Owens, former Executive Vice-President and Chief Financial Officer under Mr Scrushy, as to how HRC perpetrated these frauds, the allegations noted above – and more – were apparently all true.

PAYROLL

1. Ghost employees: This includes fictitious employees or employees who are no longer with the company. In such frauds the net pay is directed to the account of an actual person. Eric Lipkin, the former payroll manager to Bernard Madoff's company, admitted in court on June 6, 2011 that he added fake employees to the payroll file beginning in 1986. One example of a fake employee, according to Lipkin, was the son of another Madoff company employee. He also admitted to creating false reports that were sent to a clearing house for sellers and buyers of securities.

2. Falsified hours: This is more likely to occur for non-salaried employees who are required to register their work time. This likely involves collusion because an on-site employee would be used to register an absent employee as being present via the time clock.

GOVERNMENT REGULATION

The falsification of reports to regulatory agencies is fraud and can occur under different circumstances:

1. Some dedicated employees may become torn by their willingness to protect their employer's mistakes, shortcomings, flaws, or deliberate misrepresentations or deeds by covering up such incidences in regulatory reports because of their love for the organization.

2. Employees may believe that by being truthful in regulatory reports, the oversight agency could close down the organization, leaving them jobless.

3. Employees may be the subject of internal (management) pressure not to be truthful in reporting activities or certain incidents to regulatory agencies.

Frauds relating to regulatory reporting include the falsification of performance reports by public-funded agencies. Public-assistance programs funded by tax dollars are usually required to maintain a certain level of performance which is based on its success rate in helping those in need. The tax monies appropriated to such programs are often doled out through a combination and hierarchy of state, county, and city agencies in order to reach the recipients. These performance reports are used to grade the effectiveness of the government program itself and not just the agencies that awarded the funds. It is easy to see why an employee of a public-funded agency would be tempted to falsify a performance report – the agency could lose its public funding, which could result in its doors being closed and everyone losing their jobs.

MANUFACTURING

1. False counts: Machine counters might be tweaked whereby more of a product is produced than accounted for and the unaccounted product is pilfered.

2. Quality: The quality of a product can become tainted due to the use of substandard materials, illegal additives, overused dies and molds, etc.

3. Scrap theft: This would cover the theft of scrap material left over from a manufacturing process. This is especially significant when precious metals are used.

Though I had walked through the shop floor many times during a consulting assignment, I was not initially aware that the bundles of finished goods lying around were the unwanted result of inaccurate machine counts and were not being shipped out. The shop floor employees were very frustrated at the poor performance of the equipment and, in order to try and keep up with tight schedules, they force-fed raw materials into the machines, which in turn caused the count sensors to misread the actual quantity of goods manufactured. (This story is a good example of asset misappropriation related to abuse.) Bad as that was, the problem was compounded by the fact that no one was managing the wasted output being produced. Some of the workers knew that one job's incorrect output could have been another job's finished product, meaning that incorrect output from one job could be salvaged to satisfy the requirements of another job. Unfortunately there were no open communications and the supervisors did not really seem to care enough. Once this situation was revealed, it was believed that some of the incorrectly manufactured goods were being pilfered and sold at a scrap metal yard and the monies pocketed by certain shop floor supervisors and their cronies, which was likely to be one of the reasons for their laissez-faire attitudes.

COUNTERFEITING

1. The falsification of an authorized signature on a real and original document, such as a contract or a check.

2. The replication of something original – a consumer product, a financial instrument, a contract – with the intent to pass off the fake as the real thing, possibly even with an original authorized signature.

Counterfeiting is basically defined as attempting to pass off something that is duplicated without authorization as an authentic original. This is quite common in consumer products, but is also known to occur with pharmaceuticals and electronic components.

It is not just consumer products that are being counterfeited – documents and signatures can also be faked. In 2009 the once-prominent Fort Lauderdale attorney Scott Rothstein was charged over his involvement and leadership in a fraud scandal amounting to more than $1 billion that included a Ponzi scheme. In order to sway potential investors to hand over money, Mr Rothstein is alleged to have created fictitious court documents with the forged signatures of actual judges. The scheme in part relied on the use of the fake court documents which declared that Mr Rothstein's victims (his clients at the time) were to turn over cash to his law firm based on a fictitious judge's order representing a fictitious out-of-court settlement of an actual lawsuit. According to Cornell University's Legal Information Institute, Title 18, Part I, Chapter 25, Section 505 of the US Code states that:

> *Whoever forges the signature of any judge, register, or other officer of any court of the United States, or of any Territory thereof, or forges or counterfeits the seal of any such court, or knowingly concurs in using any such forged or counterfeit signature or seal, for the purpose of authenticating any proceeding or document, or tenders in evidence any such proceeding or document with a false or counterfeit signature of any such judge, register, or other officer, or a false or counterfeit seal of the court, subscribed or attached thereto, knowing such signature*

or seal to be false or counterfeit, shall be fined under this title or imprisoned not more than five years, or both.

Mr Rothstein falsified his law firm's income via the use of fake documents suggesting that his law firm was to receive money for victories in fictitious court cases. Another use for faked judicial orders could be to sway his client to pay more money than the combination of legal fees plus a settlement agreed upon by the other party, with the implication that the case went before a judge and the judge ordered a particular monetary settlement. In such an example, the attorney breaches the trust between counsel and client.

While we generally think of counterfeiting as the creation of something wholly fake based on an original, we can stretch this definition to include fake products that contain part of the original or are otherwise composed of some original parts. As reported in the February 4, 2010 edition of the *South Florida Sun Sentinel* newspaper, a state of Florida police officer assigned to the Miami area was arrested and charged with issuing fake traffic citations to drivers. The citation document (ticket) was real, as were the drivers he was assigning them to – the officer used the information of persons he had previously stopped for traffic violations. The fraud was due to the fact that the new charges were fictitious. (The components of this fraud were actual document, factual person, and faked transaction.) In some cases the charges were serious enough that the person could have lost his or her right to drive. These fraudulent citations were not signed by the drivers because they were not assigned during a traffic violation stop – they were manufactured later. It is suspected that the police officer perpetrated this fraud to boost the count of the traffic tickets he issued.

Characteristics of Supply Chain Fraud

With an understanding of the holistic nature of the supply chain, we can look at some of the characteristics of supply chain frauds.

SUPPLY CHAIN FRAUDS ARE NOT NECESSARILY A SINGLE DEPARTMENT'S PROBLEM

The links of the supply chain are all interconnected and what happens in one link can – and likely will – affect what happens in the adjoining links. When investigating supply chain frauds, we may have to consider that the cause of the problem might be shared between multiple entities, such as different departments. In addition, a fraud might be realized in one department when it in fact originated in another or in the transition from supplier to customer. Remember that gaps between the links represent areas where fraud can exist, not just within the links themselves.

A SUPPLY CHAIN FRAUD MAY MANIFEST ITSELF DIFFERENTLY FROM HOW IT ORIGINATED

Interconnected chain links mean that a fraud in one part of the supply chain can have a ripple effect throughout the whole or a part of the rest of the supply chain. We can expect the results of a fraud to appear in different ways as it moves through each of the supply chain links, because each link will be affected differently by the fraud.

> A very good example of this supply chain fraud characteristic is the tainted pet food scandal that affected the US and other countries in 2007. Melamine was added to the base pet food in China in order to boost the protein level to cover up the use of substandard ingredients. The tainted pet food was allowed to continue through the entire manufacturing and distribution process. It was only when consumers realized that their pets were getting sick – and dying – from eating certain brands of pet food that this fraud was revealed and traced to its source. In this case it would appear that the original intent of the fraud was to utilize less-than-first-quality ingredients as a substitute and reap the financial benefits. The results of this fraud were sickness and death. Thus, the fraud manifested itself as something much worse than was probably initially intended. I seriously doubt that these fraudsters conspired with the intent to kill pets belonging to citizens of Western countries; rather, it was all about greed and profit.

SUPPLY CHAIN FRAUDS ARE ABOUT MORE THAN "COOKING THE BOOKS"

Staying with our tainted pet food example, this fraud would not have been discovered by an examination of the accounting books or financial statements. Certainly, the results of this fraud negatively affected the finances of the companies involved due to the impact of product recalls, testing, lawsuits, fines, and so on. Because supply chain frauds happen in business operations, organizations must look at creating effective governance programs throughout the entire length of their supply chains. They cannot rely solely on a horizontal or vertical analysis of numbers that are the result of the business transactions to determine whether fraud happened in the past, by which time it could very well be too late.

> Horizontal and vertical analysis are key tools in the examination of financial statements and data analysis. In horizontal analysis the same data point, e.g. manufacturing output or sales revenue, is compared across multiple time periods (months, quarters, or years) to ascertain trends. In vertical analysis different data points are compared in a single time period, e.g. the comparison of sales representative commissions to overall sales revenue, to ascertain whether the percentage of one data point to the other makes business sense and is in line with expectations or industry standards.

In the detection and reduction of supply chain fraud, it is important to not begin at the end – the financial statements – but to consider them as simply one aspect of an enterprise-wide focus on the fraud. We want to prevent the infiltration of something bad into our supply chain as early as possible in order to mitigate its metamorphosis into something worse that infects an otherwise healthy organization.

SUPPLY CHAIN FRAUDS CAN BE PERPETRATED VIA VARIOUS COLLUSION MIXES

Regardless of whether an individual is acting alone or in concert with others, supply chain frauds can be committed by individuals inside the organization, outside the organization, or by a combination of internal–external collaboration. Employees might be willing participants or may actually be the victims of extortion or other influence of some kind. (And we should not be so naïve as to believe that such pressure cannot be exerted only from outside: Nefarious internal forces, especially from people in roles of authority, can also be responsible for this!) Organizations – especially those in highly competitive or secretive industries – need to be on the lookout for vulnerable employees. HR departments in such organizations should set up open lines of communication whereby an employee can discuss and get help for certain problems which less-than-reputable entities can take advantage of, and the employee should be able to discuss help without fear of job loss or embarrassment. However, HR employees may be unwilling to believe or act upon the statements of an employee against the employee's boss or an executive. This power play is a common tactic used by people in authoritative roles. HR departments must be willing to keep an open mind when staff employees come to them with concerns about the behavior of, actions by, or directions from management and executives that the employees believe are contrary to the well-being of the organization and its stakeholders, especially because certain behavior can be an indication of fraud.

Pulling off a fraud on one's own can certainly yield great gains for the fraudster. But to really make massive financial gains or to get around business barriers (standard operating procedures, software application protections, network infrastructure securities) usually requires a team. The primary difficulty in fraudsters collaborating is the issue of trust. Or, in other words, is there truly honor among thieves? Well, maybe during the commission of the fraud, but it seems as if all bets are off after everyone gets caught. The trust that conspiring fraudsters must have in the others involved must cover the following issues:

- Not stealing from the other partners by taking more than one's fair or agreed-upon share.
- The trust that none of the other partners will falter and bring attention to the fraud, bringing down the whole proverbial house of cards.
- A willingness to take the fall and not name conspirators if someone gets caught, thus protecting the others. I do not know how long this usually lasts; whenever I read an article with regard to fraud collaborators getting caught, it seems that the honor they had as thieves has been replaced by the need for self-preservation and the desire for lower penalties and reduced prison sentences.

Supply chain operations should utilize business processes and technologies that allow for a series of checks and balances to be implemented. Potential fraudsters may think twice before perpetrating disruptive behavior if their act involves them reaching out and attempting to recruit a co-conspirator, especially if their targeted collaborator's actions were verified by systematic controls or manual audits. If the targeted collaborator does not share the fraudster's desires or needs, then the targeted collaborator might turn the potential fraudster over to the authorities right away or may not be able or willing to hold up under pressure during their illicit activities or if their fraudulent behavior is discovered.

SUPPLY CHAIN FRAUDS CAN HAPPEN DOMESTICALLY, INTERNATIONALLY, OR BOTH

It is truly a global world out there. (You can quote me on that!) As supply chains have been stretched to the corners of the planet, the opportunities for fraudulent behavior have never been greater. We cannot shift our entire focus to the international arena when some of the simplest yet most damaging frauds can still occur right in the cubicle or office just down the hall. Nor can we ignore what happens outside our organization simply because it cannot be seen. A contract does not equal compliance, though complacent organizations would believe this to be so. The growing number of product recalls by the US Consumer Product Safety Commission (CPSC) shows us that, as supply chains have become globalized, there are issues that organizations are either not prepared or not willing to face.

The CPSC publishes an annual Performance and Accountability Report (PAR) which is available on its website (www.cpsc.gov). The Report cites statistics regarding the agency's performance, including those relating to voluntary recalls which are part of the agency's Fast Track Product Recall Program. According to the May 1999 version of the CPSC's Recall handbook:

> *If a company reports a potential product defect and, within 20 working days of the filing of the report, implements with CPSC a consumer-level voluntary recall that is satisfactory to the staff, the staff will not make a preliminary determination that the product contains a defect which creates a substantial product hazard.*

There were 354 voluntary recalls in 2004. In the PARs for 2005, 2006, and 2008, there were respectively 397, 471, and 563 voluntary recalls, and in each of these reports it was noted that the number of voluntary recalls was the largest in the preceding ten years. The number of voluntary recalls was 465 in 2009 and 427 in 2010. Voluntary product recalls are a subset of the total number of recalls by the agency. Products are supposed to be recalled when they "violate mandatory safety regulations or presented a substantial risk of injury or death."

Organizations are finding that, as the distance between them and their suppliers grows, it would be wiser to reduce the level of trust and increase the amount of oversight. This is not to say that all out-of-country suppliers are dishonest – quite the contrary – but just because you have outsourced a process does not mean you have outsourced the responsibility for oversight and assurance that it is performed in the correct manner: Organizations are still responsible regardless of whether they performed the work themselves or used external resources under their control and guidance.

Supply Chain Fraud Results

The results of supply chain fraud include greater risk, reduced profits, and less control. A little fraud can lead to more fraud and this can quickly spiral out of control and devour a company. A tire importer in New Jersey was forced to recall 250,000 tires manufactured overseas due to serious manufacturing flaws that resulted in safety issues. The tire importer's annual revenue was approximately $5 million per year, but the cost of the recall was estimated at $85 per tire, amounting to a total of $20 million. I do not know if that company survived, though I would imagine it would have been quite difficult under such a debt load. Had the tire flaws been caught early on, the importer could have taken steps to refuse inbound shipments. Proper sales record-keeping would have been critical in supporting the recall effort and may even have reduced the cost of the recall process.

Was this fraud by the tire importer or the tire manufacturer? It is a good question. The manufacturer is the likely culprit if the tires were knowingly manufactured to less-than-first-quality or minimum safety specifications. Was it willful negligence on the part of the importer? If so, then this incident could be labeled as fraud by the importer. However, if the tire importer innocently did not know but should have known, it still should bear responsibility. This may seem unfair, but not knowing is not an option: Organizations that import must perform acceptable due diligence and inspections as part of their system of checks and balances. What about the tire importer's employees? For example, would the salespeople be guilty of fraudulent behavior if they performed their job responsibilities adequately and were informed that they were selling safe products? From my perspective, these employees would be regarded as victims of the organization's carelessness. Not every employee at Enron (or WorldCom or Arthur Anderson) was guilty of bad behavior, yet the entire organization's employees suffered because of the corrupt behavior of a few.

Supply chain fraud should be particularly worrisome for public companies who must comply with regulatory requirements. In the US the Securities and Exchange Commission (SEC) is charged with, among other responsibilities, the protection of investors. The SEC is becoming less tolerant of public companies simply restating financial statements year after year; such behavior makes it difficult if not impossible for potential investors to determine whether to purchase a particular public company's stock and can mislead an existing investor into buying or selling either too soon or too late.

Private companies should also take note. Supply chain frauds can be just as damaging to private and public companies alike. Though private companies may not have the same level of responsibilities to their shareholders as do public companies per se, the integrity of their business can be affected just as much by supply chain frauds. Private companies and their owners looking at merger and acquisition opportunities – either as the buyer or the seller – can find their valuation put at serious risk due to fraudulent activities.

Government agencies are spending our tax dollars and must be held extra-accountable for the care of control of these monies. In the US billions of dollars each year are lost due to fraud in the medical care system and in government defense contracts alone. Politicians seem often more intent on protecting wasteful or unnecessary programs that benefit their state than on considering the well-being of the national taxpayer dollar. As a taxpayer I have found that my confidence has been more than just breached, it has been trampled over repeatedly!

The Ramifications of Supply Chain Fraud

Supply chain fraud can affect an organization in different ways and to various degrees of severity. Just as the holistic supply chain represents the breadth of an organization, the impacts from supply chain frauds can be just as encompassing and can also strike deep.

MANUFACTURING DOWNTIME

It may be necessary for machines to be taken apart for maintenance due to the introduction of substandard raw materials or because of abuse from lack of preventative care. In either case, manufacturing equipment can be damaged to the point of failing to produce first-quality output. An even worse case is when the damage is so severe that it causes the introduction of a foreign substance – such as plastic or metal from the machine itself – into the finished goods. (Consider the problems this could cause in consumable products such as food, beverages, and pharmaceuticals.) The use of substandard or tainted ingredients – especially in the production of consumables – will mean that machines will need to be cleaned and possibly stripped down to ensure the complete removal of the low-quality or foreign substances. Returning to the case of the melamine-tainted pet food, it is very likely that the manufacturing equipment required thorough cleaning after the foreign substance (melamine) was found to have been introduced into the manufacturing process.

When machines are taken offline due to such problems, production capacity becomes limited if not completely non-existent. The fulfillment of customer orders is likely to fall behind schedule, especially if the organization has implemented a build-to-order strategy and does not have an adequate safety stock of shippable finished goods readily available. Products with limited shelf lives cannot be stockpiled and stored over the long term and therefore must be produced continually. It is not economically feasible for organizations to pay for warehousing space to store an excessive inventory of products. Both of these realities result in leaner inventories which are therefore more greatly affected by supply chain disruptions.

PRODUCT RECALL

If a product becomes tainted due to problems with the manufacturing equipment or ingredients, the supplying organization may be required to directly recall or support their customers' recall of problematic products.

Product recalls involve reserve logistics in the accumulation, shipment, receipt, and (at least temporary) storage of goods. An examination may be required to determine the extent of the problem and none of the recalled stock can be destroyed until enough sampling has occurred to determine the cause and effect (it should be remembered that frauds can manifest themselves differently from how they started as they travel down the supply chain) of what went wrong. Reverse logistics can consume considerable cash and labor resources. It is very important to perform a root-cause analysis and fraud may be one of the reasons.

Depending on the product recalled, its destruction may pose its own set of difficulties. Electronics cannot just be buried or burned without preparation due to the use of hazardous metals that can release carcinogens into the groundwater or air. Food and pharmaceuticals may have to be unpacked first, resulting in extra labor charges, after which the destruction of the contents and packaging may follow different paths.

If the product can be salvaged via rework, the loss to the manufacturer can be mitigated. However, the cost of the rework effort must be weighed against the product's value and the potential revised sale price. It may not be possible to sell reworked products as new products and thus the manufacturer risks flooding the marketplace with refurbished goods. Sending remanufactured products to other countries might be a (better?) way to recapture sales revenue, but it depends on whether there was a market in the country in the first place and if its citizens can afford even a reconditioned product. Such "product dumping" is frowned upon if the reworked product's safety or effectiveness is compromised during the salvage process. After all, why should people in poorer countries not have the same product safety rights as the rest of the world?

CREDITS AND CHARGEBACKS

The supplier (who could be the manufacturer or the distributor) of a problematic product might find itself providing credits to its trading partners (customers) and/or issuing refunds direct to consumers. (This is a more likely scenario in retail but could apply to other industry segments as well.) Adding insult to injury, the supplier may face vendor compliance chargebacks for violating the terms of the supplier–customer relationship. A common feature of the US retail industry, vendor compliance chargebacks are imposed by the customer (e.g. retailers) upon suppliers who fail to comply with the dizzying landscape of supply chain requirements that are established by the customer. Vendor compliance chargebacks are supposed to penalize the supplier for the cost of the correction, but are believed to far exceed that basis at times. Therefore, the added chargebacks can compound the negative financial impact of, for example, the recall of a consumer product.

LAWSUITS

Consumers of a harmful product could sue the manufacturer, the distributor, or both for damages from injury or death. US courts may allow class-action status if it can be proved that a sufficiently large number of people were harmed to warrant such a designation. Lawsuits between trading partners, i.e. the retailer, the supplier, and the manufacturer, may also arise.

GOODWILL AND FINANCIAL LOSSES

Goodwill is generally considered to be an intangible company asset, such as the company name or product brand that gives the company greater value beyond merely its physical assets and sales figures. Consumers place great trust in certain companies and their brands, and this trust can be put to the test when products thought to be safe cause harm. A loss of reputation or brand trust can adversely affect the stock prices of publicly traded companies and can reduce the goodwill value of brand reputation.

An example of this is the Tylenol product scare of 1982, where tainted bottles of this over-the-counter medicine were shown to have been tainted with cyanide and caused the death of seven people in the Chicago area. Johnson & Johnson and its subsidiary at the time, McNeil Consumer Products, not only had to deal with the recall but also the public relations crisis that made people question whether the company's other products were safe too.

In 2010 Toyota faced the effects of a recall of millions of vehicles worldwide due to problematic accelerator pedals and intrusive floor mats. There was criticism in the news, and even the Toyota president, a grandson of the founder, stated that the company waited too long to report the problems and take action. The Toyota brand is/was synonymous with quality and now this reputation is under question and may take years to recover.

In another case, while the damage done was not very funny, I do have to laugh – as do my audiences when I use this example during speaking engagements – at how one company handled its apology to consumers relating to a tainted peanut butter spread product. The company offered consumers who were made ill a coupon for a free jar of – yes that's right – its own brand of peanut butter spread! (The coupon offering was after the recall and replenishment of new jars of peanut butter to store shelves.) My audiences and I find this a bit underwhelming: Is this all that the company thought of its consumers? What does the limited extent of its apology say about how this company cares about its actions and the problems that it caused? As a consumer, how much trust would I now have in this company and its products? Or did this company simply rely on the fact that consumers have very short memory spans and would soon forgive and forget? (To answer the last question, I am sure that consumers did just that not too long after the incident.) As I tell my audiences, if I were the CEO of that company, I would have at least included coupons for free bread and jelly too as a sign of good faith!

Losses in brand trust can cause consumers to switch to a competitor's product. This could result in irreversible losses in reputation and future sales to the company whose products caused harm to consumers. An organization's stock prices could take a hit, possibly affecting its ability to secure loans or lines of credit, causing projects to be delayed or cancelled due to lack of financing, and therefore affecting the employment of an untold number of people, not just employees but suppliers and contractors as well.

I am left to wonder if some (most?) organizations rely too much on short-term consumer memories and strategic marketing campaigns versus investing in the necessary people, processes, and technology to avoid problematic product incidences in the first place. Certainly, accidents will happen and honest mistakes will occur. Yet I cannot help but believe that unnecessary sacrifices are being made for the good of short-term profits over long-term responsibilities and that some organizations are more focused on risk management than customer service. Organizations may begin to foster a culture of bad behavior that is more focused on damage control than product safety, calculating the return on investment of legal settlements versus the implementation of quality control programs, technology, and auditing measures. This is an organization attitude that starts at the top with executive management and is passed downward through the ranks of management and rank-and-file employees. In a sense this becomes the personality of the company.

DAMAGE CONTROL AND DELAYED RELEASES

Companies whose products cause harm to customers will likely find themselves in the unenviable position of having to allocate funds not just for the recall efforts but for damage control advertising as well. Notifications about product recalls will typically be run in major newspapers and magazines, but they may also appear on television.

During this crisis period, any releases for new or improved products might have to be delayed, including sponsorships and advertising spots on billboards, in printed media, on radio, and on television. It is likely that consumers will not want to hear about what's new and improved about a company's products while the old product is causing harm and is being recalled. Even if the opportunity is not pounced upon by competitors, consumers are nonetheless likely to look elsewhere for a compatible product and may make the switch permanent.

In the Toyota recall case, the auto manufacturer ran damage control advertisements in newspapers that stated its commitment to quality and explained its dedication to carrying out speedy repairs. (I saw such an advertisement in the February 2, 2010 edition of the *South Florida Sun Sentinel* newspaper.) Toyota ceased production of the problematic models during this crisis, leaving dealers with depleted inventories of vehicles to sell. Aside from Toyota's corporate headaches, this crisis had the real potential to force independent dealers to cancel their own advertising during the recall effort, exacerbating the costs of the recall through the dealer supply chain.

REGULATORY INVESTIGATIONS

The salt in the open wound of a supply chain fraud with critical consequences may be government regulatory agency investigations to determine the cause of the problem, what laws might have been broken, and if the organization should have had necessary – if not legally required – processes in place to prevent it from happening in the first place. Fines, lawsuits, and criminal arrests of key executives (for which the company may be required to pay the defense based on the terms of each executive's contract) are all possible outcomes of regulatory investigations and place considerable financial burdens on the organization.

The Rights of the Customer

The first ten amendments of the US Constitution are also known as the Bill of Rights. These include foundational concepts like the freedom of the press, free speech, freedom from religious persecution, the right to bear arms (controversial but generally upheld), and the right against unreasonable search and seizure. In essence these are the basics upon which all other rights are considered. From what we have examined thus far, in looking at the holistic supply chain which is composed of suppliers and customers at each link, is there a similar set of basic rights that customers in a supply chain can expect as they interact with their respective suppliers? Yes – I believe there is.

I found the following customer's bill of rights in the February 2005 issue of *Inbound Logistics* magazine. These state that the customer should receive the:

- right *product* in the
- right *quantity* from the
- right *source* to the
- right *destination* in the
- right *condition* at the

- right *time* with the
- right *documentation* for the
- right *cost.*

In essence, isn't this what all organizations should be striving to achieve, not just for their external customers but also for their *internal* customers as well? Certainly, this is applicable to a manufacturing and/or distribution organization that goes through the process of order preparation, assembly, and shipping. But it is equally relevant to a host of other industry sectors, such as banking and financial firms in the processing of documents for loans, investments, lines of credit, and other contractual agreements. Similarly, a hospital patient should expect the same in the receipt of pharmaceuticals, tests, and medical procedures all geared toward discovery and cure. We can look at fraud detection and reduction as a guarantee of the protection of the basic rights of an organization's customers – both internal and external. Engaging employees in fraud detection and reduction may be easier if it is made part of an operations excellence plan by the organization. The customer's bill of rights shines the light on key supply chain points where fraudulent activity can occur.

Note that the customer's bill of rights in this context applies to customers who are potential victims of fraudulent behavior, not customers who are perpetrators of, collaborators in, or co-conspirators of fraud.

CUSTOMERS DESERVE THE RIGHT PRODUCT

What was supposed to be delivered to the customer?:

- Did the customer order a pair of shoes but was shipped a pair of pants?
- Was the customer supposed to receive a certain quantity of raw material samples for inspection?
- Was the right form used to begin a loan application?
- Did the pharmaceuticals delivered to the patient match those that were prescribed or were substitutions or compromised products with reduced effectiveness used instead?

Fraudulent behavior can result in customers not receiving what it is they should be receiving, and in some cases the results can be deadly, as in the case of pharmaceuticals.

IN THE RIGHT QUANTITY

All in all, if the customer is receiving an incorrect quantity of something, the question that has to be asked is "why?" If a customer orders two shirts and receives 20, that is a problem. If a customer was approved for a $10,000 loan but receives $100,000, that is a problem too – and quite possibly a bigger one at that. If 100 of any item were ordered but only ten were received, where are the other 90? Fraudulent behavior may be involved in the processing of inaccurate delivery quantities and may be an indicator of returns fraud.

It is fraud if a hospital worker reduces the recommended dose of a pharmaceutical to a patient for ulterior motives. For example, a patient is supposed to receive two 250 mg tablets of pain reliever every four hours, but a hospital worker only gives the patient one tablet, pocketing the other one for his or her own use or sale later. It is fraud if an organization deliberately reduces the quantity of a raw material or ingredient that is supposed to be included in a product. It is also fraud if an organization deliberately allows a product or service to be sold whereby the customer is paying for one amount or service level but is receiving less.

FROM THE RIGHT SOURCE

It is important to ensure that the right department is acting as the supplier in the internal supply chain relationships. Remember that the handoff from a supplier to a customer should be complete. This means that the right supplier should have absolute and unshared authority to perform the move. This also means that the right supplier has the burden of this responsibility, which should force the supplier to ensure that the move is the right thing to do, lest it suffer the consequences of its actions. These kinds of operational deficiencies open the way for fraudulent behavior performed by otherwise honest employees to compensate for a failed business process. Employees may know that something should not be moved from Point A to Point B, but they are just told to do their jobs and are criticized when they do not. If the employee does not have the authority to prevent a move, or if the employee is at risk of losing his or her job from being the continual bearer of bad news, he or she is likely to have more incentive to remain quiet even when he or she knows about a potentially liable situation.

"From the right source" also means knowing your supplier. Hospitals should ensure that their pharmaceutical distributors are sourcing products from validated organizations that have passed appropriate regulatory screening procedures. US retailers have strict guidelines that prohibit their suppliers from using manufacturing facilities – their own or those of their contractors – that engage in child labor practices.

TO THE RIGHT DESTINATION

It is important to ensure that the right supplier is in fact handing off whatever is moving through the supply chain to the right customer. I have lost count the number of times I have been at a client site and someone in a department asks: "Why are we getting this when we do not have responsibility for handling it?" Pass-through departments are the cause of disruptions and result in inefficiencies. If something was forced through an abnormal supply chain direction, we should ask "why?" and consider that fraud may be involved in the unnecessary interim step. If a supply chain is unnecessarily complex, it opens the way to a greater potential to falling victim to illicit behavior.

In a hospital this equates to ensuring that the right patient is receiving the prescribed treatments, tests, and medications. Two patients with the same name or similar identification numbers can cause confusion. Even law enforcement departments have been known to confuse the identities of people and address locations. As such, it is necessary to ensure that shipments of products are not rerouted by unauthorized persons to unauthorized destinations. Falsification of a destination – whether a person or a place – via the use of a similar name to the actual person or place is one way to perpetrate this kind of fraud.

Let us recall the example of the public official who set up a dummy services company and stole $500,000 from his employer – a city – before he was caught. The perpetrator directed that he should receive the invoices sent from his dummy company and that he would authorize their payment. If this person's normal job role did not include invoice authorization, then – aside from his directive – why were invoices passed to him for payment approval? This could have been a warning that something was not right with the redirection/misdirection of that particular (fraudulent) vendor's invoices.

IN THE RIGHT CONDITION

Remember: When negligence is willful, it is done on purpose. Employees who willfully disregard how they handle anything – from hazardous materials to apparel to documents – may be breaching the confidence their employer has in them and thus could be guilty of fraudulent behavior. When ensuring that a box is not crushed, that a document covers all the necessary terms and conditions, or that information is entered accurately into a software application, employees have a responsibility to deliver what they are carrying forward – physically and technologically – in the right condition.

It is also important to bear in mind that damaged goods still have value. Deliberately damaging a product to render it unsellable and then absconding with the product for one's own purposes would be indicative of fraudulent behavior. This fraud can happen during the returns process or when items are initially picked for shipment.

Food, beverages and pharmaceuticals – including their respective basic ingredients – are examples of products with limited shelf lives and expiry dates. The sale or use of a consumable item past its recommended "best use" date may result in illness or further medical complications due to the use of a weakened medication.

AT THE RIGHT TIME

Something that arrives too early or too late can be disruptive to supply chain operations. "Early" and "late" can represent minutes, hours, days, weeks, or months. Inventory that arrives too early can force an organization to incur extra storage costs and could reduce cash-in-hand to pay the supplier invoice. Inventory that arrives too late may require expedited shipping and could delay manufacturing or order fulfillment. These may be indicators of someone trying to make a certain supplier look either good or bad in the eyes of the organization.

Information upon which we are going to base operational decisions or investment strategies must be delivered to us appropriately. Information that arrives too early may exclude data from a certain timeframe. Information – such as financial statements or analytical reports – that arrives too late may have been critical in making an important decision, perhaps even causing a decision to be delayed altogether. Are delays or deceptions being done on purpose? If so, there could be fraudulent behavior afoot.

WITH THE RIGHT DOCUMENTATION

Organizations sending and receiving shipments will require manifests, bills of lading, and packing lists to verify the contents and comply with various transportation regulations. The import or export of goods requires additional documentation to comply with the laws of various national governments. Regardless of whether the documentation is in paper or electronic form, its necessity remains intact.

Different industries will have different supply chains and therefore different documentation requirements. When purchasing a home in the US, there is a mountain of documentation that is required during the process, even documents where the buyer attests that he or she has received other documents. So when we state "with the right documentation," it means not only original or source documents but supporting documents as well. The "right" documentation can also be a reflection of the documentation being accurate and not containing fraudulent statements or amounts. Mortgage frauds will often include misstatements about the income of the buyer or the value of the property being purchased.

FOR THE RIGHT COST

Did the supplier deliver what it was supposed to at a reasonable cost? Did willful negligence result in excessive costs being borne by the customer? Were prices raised without customer knowledge or approval, possibly in conflict with stated terms? Were there add-on fees for products or services that were not needed? Have suppliers conspired to fix prices on certain ingredients? In each case fraudulent activity may be at the root of higher than expected costs for the customer. All industries – manufacturing, legal, health care, consulting, and more – are ones where unethical suppliers may look to line their pockets with fraudulently billed – and collected – money.

Fraud in the supply chain can cause disruptions that rob our customers of these basic rights. As organizations move toward the establishment of policies and procedures, the implementation of technologies, and the closing of supply chain gaps with the intent of fraud detection and reduction, this process should be performed with the theme of protecting the basic rights of their customers, both external and internal. Fraud detection and reduction programs should be mixed with operational excellence plans.

Detecting fraud requires forming the basis of what is normal behavior. Deviations from the baseline – and the significance of the deviation is subject to definition depending on the supply chain or the position within the supply chain – are the alerts that bring our attention to something that is not quite right. These aberrations can be representative of bad behavior that causes disruptions, the exceptions to the norm. Whether the bad behavior is fraud or laziness is still to be discovered, but without the formation of a behavioral baseline, we cannot set alerts for deviant behavior. I think that there is also value in examining behavior that is consistently perfect because it is here where we may find fraudulent activities being covered up, albeit covered a little too well. Students who hack their school's computer systems and change grades cause suspicion to be raised when an average student's grade is suddenly excellent. This is likely to be a combination of greed and ego. If fraudsters want to reduce the chances of getting caught, they have to cover whatever tracks they may have left and make it as unclear as possible that a fraud has occurred at all.

PART II
Detecting Supply Chain Fraud

This section will dive into the detail and will examine the supply chain frauds that can exist at each link of the inbound and outbound supply chains and in the gaps between each link. Remember: This book is not a how-to guide for perpetrating fraud. (So don't try this at home, at work, or at all!) I will review motives, collusion possibilities, conditions (such as lapses in procedures and gaps in software applications), and will detail step by step how different supply chain frauds can be accomplished. Detection strategies and how some readily available technologies and supply chain methodologies – including metrics, key performance indicators, and scorecards – can be used to help detect supply chain fraud will round out this part of the book.

It is likely that fraudsters who must recruit collaborators to engage in their illicit activities will have to provide some type of remuneration or incentive to their (willing or unwilling) participants. This can be in the form of a bribe or a kickback; incentives can include all-expenses-paid vacation getaways, show tickets, or jewelry. There can be a very fine line between what is a bribe or kickback versus what is a gift, and the gray area is sometimes exploited when the fraudsters are caught.

A bribe is when something of value is provided to someone before that person makes a decision or takes an action. The intention of the person providing the bribe is to alter the behavior of a person in a position of authority or other decision-making capability. A kickback is when something of value is provided to someone after that person makes a decision or takes an action. The intent of the kickback – like the bribe – is the same: To alter a person's behavior in favor of the person providing the something of value. It is generally easier for a kickback to be based on the value of the outcome because it occurs after the fact.

The differences between a gift versus a bribe or kickback are as follows:

a) A gift is an expression of thanks and is not illegal like a bribe or a kickback.

b) Unlike a bribe or kickback, a gift is not intended to alter a person's behavior so as to force a dubious decision in favor of the gift-giver.

This can be a tricky path to navigate. Could a gift be of such a great value that it would sway the recipient's decision until the next year's holidays? (Yes, and this is why some organizations have strict gift-giving policies.) The difference is that a gift is not illegal; it is an expression of thanks and is not intended to sway someone's decision away from what is right. Bribes and kickbacks are illegal and intended to alter someone's behavior away from what is right and toward what is favorable for the person providing the bribe or kickback. The subtle differences are less confused when clearly defined in an organization's employee manual, through employee ethical behavior training, and in vendor compliance guidelines documentation.

An incentive is not always something that is nice: Sometimes a fraudster's incentive is the withholding of certain information which, if exposed, could compromise another person's standing within the community or organization. The fraudster could be forcing the collaborator to assist in the illicit behavior due to blackmail, hence the collaborators could be innocent victims. The threat of job loss or exposing a weakness (e.g. gambling debts, drug abuse, an illicit affair with a co-worker or someone outside the organization) could be used as leverage to force a person to choose between perpetrating fraud on behalf of the fraudster or facing the consequences of termination or having his or her dark secrets made public, causing embarrassment and perhaps isolation from the community (e.g. friends and co-workers). Weighing fraud against an organization versus personal humility, it is not unreasonable to imagine that a person could be coerced into perpetrating fraud when faced with the risk of loss of social status. A fraudster could produce evidence to a collaborator of his or her unknowing or unwilling participation in a fraud scheme and thus use this as leverage to force the collaborator to assist in a continuation of the fraud scheme under threat of the evidence being made public. Fraud conspirators who have schemed together and agreed to act as a team to engage in fraudulent activities will have decided how to split their take during their own negotiation process. The reasons for perpetrating the fraud are not likely to vary much, mostly relating either to greed or need. Fraud motives do not change much, but the methods do.

3

Catalyst Event Frauds

Inasmuch as each link of the supply chain becomes active due to the firing of a trigger event, there must be an initial event to start the chain reaction. What is the initial action that is the catalyst for the other actions in the supply chain? It is the organization's need to acquire something.

First, we need to review a few key definitions related to inventory control. I do not want to delve too deeply into inventory control theory, nor do I want to get caught up in semantics. I think the following definitions are accurate and acceptable for this discussion, even if they are a little oversimplified:

- On-hand quantity: The current amount of available product inventory.
- Safety stock quantity: The minimum desired amount of available product inventory; the minimum desired level of the on-hand quantity balance.
- Re-order quantity: The standard amount of replenishment product to re-order.
- Re-order point: The amount of product inventory (the balance of the on-hand quantity) that will cause a replenishment order (for the re-order quantity) to be issued. The re-order point is generally considered to be the amount of inventory that, based on consumption and supplier lead time, will allow an organization to maintain at or above the safety stock level between replenishment orders.

In supply chains such as manufacturing, distribution, and retail, the initial trigger event is most likely the situation when the on-hand quantity of something (a raw material, component, or finished good) falls below the established re-order point. Examples include shampoo on a retail store shelf, medical supplies (from bandages to pharmaceuticals) in a hospital storage room, and plastic used in injection molding. This could occur in inventory control systems, warehouse management systems (WMS), vendor management inventory (VMI) partnerships, ERP systems, and point-of-sale (POS) applications.

Data fields like the on-hand quantity are typically maintained by the application software itself based on transactions (e.g. inventory adjustments, purchase order receipts, sales order fulfillments) that would increase or decrease the quantity and are therefore not usually able to be directly maintained manually. Sometimes specific user security must be enabled to adjust a "living" (software-maintained) data field. If a user is able to manually adjust a living data field, I would certainly hope that the software has logging capabilities to capture a non-transactional data field change and that this feature is turned on. If the answer is "no" in either case, then it is easy for the software to be the subject of fraudulent adjustments that cannot be tracked back to anyone.

In some software applications, setting different supplier lead times by product might be possible, and thus the re-order point might be a living data field, calculated based on the lead time, consumption (average or actual and either of which could also be a living data field) and safety stock. For this discussion I will assume that the re-order point is a living data field and that authorized users can affect product-specific supplier lead time and the corresponding product safety stock quantity. I will also proceed on the assumption that product consumption is maintained by the software and is not subject to manual adjustments.

Buyers, planners, inventory control personnel, and purchasing agents represent job roles that are likely to have the ability to change safety stock quantities, supplier product lead times, and re-order quantities as part of their responsibilities. Valid reasons for changing these data fields include reacting to forecasts, market trends, advertising promotions, production capacity, out-of-date product disposal, and special events. The reason for the fraud will determine whether the fraud is about holding more inventory than necessary or less inventory than required.

Holding More Inventory than Necessary

Imagine this scenario: An existing supplier, supplier representative, or distributor (Sam) engages in the following conversation with Paul, an organization's purchasing person:

Sam: Hi Paul – thank you for taking the time to see me today. You know I always like to personally check in on my favorite customers and make sure we're delivering quality product on time.

Paul: It's always nice to see you Sam. Sure – I think your company's performance is just fine.

Sam: That's great to hear. How's your family doing?

Paul: They're fine but you know how it is. My son's going to need braces soon and we're looking at different universities for my daughter. It seems that higher education also means higher prices than when we were that age! She's been accepted by some very well-known and high-quality institutions but with my wife laid off from her job I really hate the thought of having to compromise on her university education, but we might not have a choice even with some scholarship money she's qualified for.

Sam: Gee – I hear you Paul. Life – especially with a family – sure is tough these days. I'm in a similar situation with my family, which is also why I wanted to talk with you today. With the bad economy my customers are buying less so it's really hurting my sales figures. I was wondering if you could find a way to increase your sales volume a little bit; it would really help me and the slight increase wouldn't require your organization to hold much more inventory. I'd be very grateful to you especially when the end-of-year holidays come around.

What defines a kickback versus a thank-you gift can be tricky. The outright giving of cash probably more resembles a kickback than a gift. Giving tickets to a show or sporting event may be an acceptable business "thank you" gift. The use of vacation homes, airline tickets, hotel or spa stays, electronics and jewelry may be more closely associated with items used in kickbacks. The distinction can be problematic when left undefined. This is why organizations need to clearly define gift-giving policies and provide this information to employees and suppliers alike. Here the supplier is asking his customer to increase the order quantity beyond what is needed and is suggesting a kickback for doing so. The reason is pretty straightforward: The supplier (whether the supplier's salesperson, an independent sales representative of the supplier, or a distributor) is looking for more sales or more commissions depending on the job role. We do not truly know if Sam's family pressures are the same as Paul's, so we do not really know what Sam's motivations are. (Sam could have someone on the side he is having an affair with, he could have a drug addiction or gambling problem, be feeling the desire for a fancier vehicle or boat to keep up with the neighbors, is jealous of a more successful sales representative at his company, is feeling some pressure from his boss to increase sales ... the possible reasons are many.) The notion of a kickback is certainly representative of fraud. Hopefully Paul is an ethical person and will politely refuse Sam's offer. Paul should report Sam's behavior to his superiors and his HR department at his organization. It is possible that Sam was elicited by Paul's organization to test Paul's honesty and ethics, but it is more likely that Sam was acting on his own for his own benefit. If nothing else, Paul may simply be too fearful of getting caught and thus being dismissed from his organization, leaving his family without their primary income-producer, and will not be involved in Sam's scheme.

If Paul goes along with this, he either does not see the full ramifications to his organization or he does not care enough versus his own needs. By increasing the safety stock quantity, supplier lead time, or re-order quantity, Paul's organization is forced to hold more inventory in stock than is required. The results of this would include the following:

1. Less space for other inventory with the possibility of additional space acquisition being required.

2. More time required for cycle counts and full physical inventory.

3. Less cash in hand for the organization because more inventory must be paid for.

4. Greater risk of loss if the inventory becomes out of date, is stolen, or is lost due to disasters such as fire or flood.

5. Higher insurance premiums due to owning more inventory with higher claims if there is a disaster.

Holding Less Inventory than Required

Let's change the scenario and imagine that a competitor of Sam's (called Carl) is having a meeting with Paul in an attempt to secure some business:

Carl: Hi Paul – thank you very much for seeing me today. I appreciate your time.

Paul: Not a problem Carl. I should let you know that we are pretty happy with our current supplier relationships but I guess it doesn't hurt to consider other sources.

Carl: I can really appreciate how valuable close relationships are with key suppliers and I hope that once you learn more about my company, we can be one of them. I see from the pictures around the office you're a family man. Do you mind if I ask how old your children are?

Paul: My son is 10 and my daughter is 17.

Carl: My wife and I have three children between the ages of 6 and 12 – two older boys and a daughter. The middle one is about ready for braces we think.

Paul: Yes, my son just got them.

Carl: I'll bet your daughter is about ready for university, right?

Paul: Yes we're in the process of picking one now.

[Carl gives Paul the sales pitch for his company's line of products and services.]

Carl: You know Paul, I'll bet our prices and products are just as competitive – if not more so – than your current suppliers. I know it can be tough switching suppliers, but I can offer some financial incentives to you based on a percentage of the orders we get from your company – kind of like a commission plan if you will.

Paul: You mean like gift baskets for the holidays?

Carl: Well I think I can do better than that Paul. You know, there are direct financial incentives like cash and indirect financial incentives like sports or show tickets, maybe a spa day for your wife to enjoy, those kinds of things. I'll let you choose whatever works best for you.

> *[Paul knows the extra cash would be very useful and his wife would appreciate a show once in a while, something their family budget does not currently allow. But Paul is not sure how to divert some of his organization's business to Carl's company. He is a little upset that none of the current suppliers has ever offered these kinds of financial incentives for just buying what his organization would normally buy anyway, especially after all the years of good business Paul has sent their way.]*
>
> *Paul: Well it looks good Carl – I'll think about how we can do some business together and if your products and prices are as good as you say – and the rewards are good enough – we might be able to do a lot of business together.*

Like the prior scenario, the financial incentives – veiled as a commission – Carl was offering Paul are really just a kickback. Carl was wise in scanning around Paul's office to find avenues that might make the introduction of such valuable rewards more desirable by appealing to some of Paul's weaknesses or pressures. If Paul makes the decision to succumb to Carl's rewards for sending business to him, how could Paul successfully accomplish this, since his employer is already happy with its current suppliers? If prices and performance are the same, what reason could Paul use to get Carl's company on board as a supplier?

One valid business reason for adding a comparable supplier – beyond just price and performance – is simply good risk management. Paul could use the argument that too much of the organization's business was in the hands of a single supplier and that it would mitigate risk to bring another supplier on board. This would be smart thinking, but it could open up a whole proverbial can of worms: Paul's suggestion might cause his employer to go through a full-scale process of reviewing and selecting alternate suppliers, leaving Carl's company possibly out of the running, especially if Carl's company is not in the same league in terms of product, price, or performance as its competition.

Paul could go to his current supplier – Sam – and hint at some financial incentives for all the business Paul sends his way. But there are problems with this course of action:

a) Paul may not be have the authority to select or change suppliers and thus he may have no leverage to make such a suggestion knowing that he could not switch from using Sam's company;

b) the supplier may report Paul to his employer for suggesting the perpetration of such a kickback-type fraud. No … Paul is going to have to think of another way to get Carl's company in the door.

By decreasing the safety stock quantity, supplier lead time, or re-order quantity, Paul could force inventory out-of-stock situations and make it look like the current supplier's performance has taken a dive. Replenishment orders will not be sufficient or timely to cover the product balance gap. After this crisis occurs a few times, Paul can be the hero and can introduce a new supplier – Carl's company – to save the day. Since Paul is the main contact for the current supplier, he can falsely state that he has been investigating the current supplier's failure with no results. He is in a good position because his company is not likely to suspect its own employee at the outset, and this fraud would not be revealed unless someone else at Paul's company became involved in order to investigate the sudden problems with the current supplier. Once Carl's company is set up and inserted as a supplier used to cover the crises, Paul can look to divert more business to Carl's company over time.

The ramifications to Paul's company from the perpetration of this fraud include the following:

a) If the product was a finished good on a store shelf (physical or virtual), the stock-outs would result in lost retail sales.

b) If the product was a component or raw material, the stock-outs would result in delayed manufacturing or assembly and therefore delayed sales. To compensate, Paul's company may have been required to expedite the shipment of the finished goods to its customers, therefore incurring unnecessary increased operating costs. Assembly or manufacturing idle time may have resulted in a temporary work stoppage and lost wages for hourly employees.

Covering Your Tracks

After either fraud is perpetrated, Paul should reset the altered data back to its original values, thus covering his tracks, even if only to a degree. If the software that Paul's company is running has full logging capabilities and this feature has been activated, there is a chance that the software has recorded Paul's data manipulations. Of course, someone at Paul's company must suspect Paul's scheme and have the wherewithal and/or login access to investigate this. If the software does not record such direct data field management and/or the logging features of the software are not activated, then this is analogous to Paul leaving no discernible tracks of his fraud.

Which Data Should Be Manipulated?

Between the supplier lead time, safety stock, and re-order quantity, which data field would have been the better – or best – choice for Paul to manipulate to perpetrate either fraud scheme?

Remember that a fraudster does not want to get caught, at least not until the psychological stress of guilt takes over, which may never occur. Therefore, the fraud perpetrator wants to alert as few people as possible to the fact that illicit activity is occurring. Altering the supplier lead time or safety stock affects the replenishment order cycle and thus the invoice frequency. Altering the re-order quantity affects the replenishment order quantity and thus the invoice amount.

In my opinion it is more obvious to audit for quantities (amounts) than it is for the frequency of which something occurs; value differences tend to stand out more than variations in frequency. Minor frequency variations, e.g. ordering every five weeks instead of the usual six-week cycle, might be especially hard to spot if there is a natural tendency for the frequency to shift a little one way or the other to compensate for holidays, for example. Such order frequency changes would be more noticeable over a longer period of time versus sudden spikes in order quantity and thus the invoice amount. This means that the fraud could go undetected for a longer period of time and that the detection itself requires more audit effort; someone would probably have to specifically look at the purchase order cycles and perform a comparative analysis over several months, if not longer, on the sales order data. Higher re-order quantities will increase supplier invoice amounts, while lower re-order quantities will decrease supplier invoice amounts. These are easier differences to catch in an audit and to require an explanation for than the frequency of something happening and why. A sharp-eyed accounts payable employee may bring the invoice spikes to light.

Leaving the re-order quantity alone can help to avoid running into software settings that could alert others of the fraud. Some enterprise software applications may enable limits to be established per user per purchase order, whereby any amount over the user limit requires management approval. Increasing the re-order quantity could result in purchase orders in excess of the user's limit and would thus raise a red flag. A more sophisticated software application might also enable limits to be set – again, per user – based on purchases over a time period, e.g. a calendar month or fiscal quarter. In this case incrementing either the re-order quantity or safety stock could cause a red flag to be raised if the overall purchases over a time period were in excess of the user limit. Fraud perpetration is about risk management too, and a smart fraudster will consider his or her chances of getting caught before engaging in the illicit activity, as well as whether the fraud is focused on long-term or short-term results.

As reported in the November 21, 2008 edition of the *South Florida Sun Sentinel*, a then former supervisor at the State of Florida Department of Children and Families pleaded guilty in court to stealing approximately $1.54 million of taxpayers' money. The fraudster had worked for this agency for 20 years and knew how to get around the various manual and automatic checks and balances. There were 1,725 separate instances of embezzlement, with none of the individual amounts greater than $900; sometimes there were several disbursements in a given day with none apparently greater than the $900 limit. It was estimated that the money stolen by the supervisor could have helped 8,810 families in need for one full month. The former employee, who used the money to purchase luxury vehicles and to gamble, was sentenced to 17 years in jail and was ordered to pay restitution totaling $2.5 million.

This is a very good example of how an employee was able to perpetrate fraud by flying under the radar: $900 would seem to be the upper per-incident limit that this employee could authorize without further approval. There does not seem to have been any (software) restrictions on the number of or dollar amount of accumulated disbursements over a time period, thus making the fraud easier to get away with over a long period of time.

Push versus Pull Systems

Catalyst event frauds can occur in both push and pull acquisition business models. In a pull model, purchase orders are generated by a warehouse management system or point of sales system and are subsequently submitted to suppliers for replenishment of what is needed (raw materials, components, or finished goods). In a push model – such as vendor-managed inventory – the supplier is advised of product usage and replenishes the customer based on need and pre-negotiated quantities, prices, and timeframes.

Perpetrating a catalyst event fraud in a push model can be achieved when someone on the customer side alters the quantity consumed in the sales advice or similar information that is then sent to the supplier. The supplier either sends too much or too little for the replenishment quantity as it looks to maintain the customer's inventory levels based on the contract parameters. If the customer is not keeping a close watch on the replenishment, the supplier could be perpetrating a fraud scheme that involves sending too much product at a higher quantity than what was required based on the replenishment agreement. In a customer-supplier collusion where the supplier is sending more product than is required, the excess inventory could be stolen by the individual on the customer side perpetrating this fraud. In small increments this may go unnoticed by the customer organization who is left with holding more inventory and less cash because of it.

4

Purchase Order Frauds

Catalyst event fraud showed how events leading up to the generation of a purchase order can be skewed to favorably benefit an inside person and outside entity regardless of whether there is an existing or potential relationship. Frauds against the purchase order itself can be varied and aided by both real and fictitious entities. Easier to perpetrate via wholly manual processes, too many organizations have discovered that the use of enterprise software does not prohibit this fraud from happening again and again.

Purchase Orders to Real Suppliers

In this scenario, an unscrupulous employee in a purchasing role could create fake or unauthorized purchase orders directed at real suppliers. The supplier would ship the goods to the customer as normal and would provide a corresponding invoice afterwards. The fraudster might intercept the inbound shipment at some point and steal the goods. Shipment interception in transit would probably not be possible, so it would be more practical to pilfer the goods once they have been received into the facility (e.g. a store, hospital, or warehouse).

If the goods are to be stolen before being inserted into the organization's inventory, it is likely that there is collaboration between the purchasing person and someone in a receiving role. If the goods are to be stolen after integration to inventory, this could be a one-person job; however, access to inventory from someone who should not be or is not normally in the inventory area might raise some eyebrows from the regular staff and might result in the fraudster's plans being revealed.

This particular fraud probably has a short lifespan because the fraudster's organization – the buying party – will be receiving invoices from real suppliers. It might take anywhere from one to a few months before someone (hopefully) becomes suspicious as to why quantities of goods are being repeatedly ordered when there is no actual demand or why this inventory is not visible or available upon inspection. Inventory checks or cycle counts would help to detect this fraudulent activity.

Purchase Orders to Fake Suppliers

In this scenario, a fictitious supplier is established, but the supplier is in reality the fraudster or a family member or friend. Purchase orders are submitted to the fictitious supplier who then subsequently submits an invoice for payment. Purchase orders for services are likely to raise fewer eyebrows – at least initially – because it is generally harder to track intangible services than tangible goods. The fraudster issuing the purchase order to the fake service supplier may be the same person approving the fictitious service work product and fake supplier invoice. The results of the fictitious work product may find their way into the fraudster's management report, thus not requiring the falsification of a fake service report itself. (The fraudster could state the project updates were done via phone calls and the contents of the phone calls were included in management reports.) This fraud may entail payments going to a fake supplier (in reality the fraudster) during phases of a fake services agreement, giving the fraudster time to create fictitious work product documentation.

The fake supplier could, however, be providing actual goods and services, but the fraud might be that the supplier was not vetted and properly sourced: Nepotism is generally frowned upon by most organizations, especially when not revealed upfront. The fictitious supplier may be providing real – and needed – supplies, but at a higher price than necessary than if the organization went through a proper procurement process. A friend or relative may be engaged to provide services that he or she is not actually qualified to perform. It continues to surprise me when I read an article about this type of fraudulent activity and learn that the fraudster used his or her home address for the address of the fake supplier. Cross-checking employee addresses to vendor addresses is a useful audit in helping to expose this type of fraud.

Order Stuffing

As we saw in the discussion of catalyst event fraud, the frequency of real or fake purchase orders to real or fake suppliers can be increased by manipulating data. The frequency of purchase orders to suppliers can also be increased manually merely by sending more purchase orders and thus without the need for data adjustments. If there is no enterprise software application or the various operations functions are distinct – even from a centralized software standpoint – this fraud can be relatively simple to perpetrate. The chances of someone in another area – accounting or the warehouse – saying something that raises a red flag becomes less likely. This is mostly because each person will be focused on his or her own area of concern and will not necessarily see the "big picture." Sometimes purchasing professionals carry enough clout and are able to keep others at a distance. The fraudster's goal is likely to be unchanged: To purloin the excess products for his or her own profit or to benefit from kickbacks.

Part Substitution

Organizations may source products from multiple suppliers who provide comparable goods. This is a good supplier risk management strategy and helps to keep prices competitively reasonable. Medical supplies, raw materials, electronic components, nuts and bolts – the list is nearly endless, as there are no shortages of suppliers competing for the same business. If a supplier provides a limited number of products the ability to swap items on a purchase order delivery are more limited than if the supplier had a more diverse inventory of products. The distributor might be limited in its ability to swap one manufacturer's product for another to fill a purchase order depending on the industry (e.g. medical/pharmaceutical, aeronautical, military) where detailed specifications differentiating seemingly similar products are not uncommon.

The following is an interesting twist on part substitution fraud. This story highlights that organizations need to be on the lookout for used parts substituted for new parts.

As reported in the March 27, 2010 edition of the *South Florida Sun Sentinel* newspaper, after being accused of running a $40 million boiler-room operation in New York, the owner posted bond and headed down to Palm Beach County, Florida, where he and his co-conspirators continued their lifestyle based on criminal activities. (A boiler-room operation is one where high-pressure sales calls are made to susceptible people – usually senior citizens – for the purposes of selling useless, worthless, or non-existent products and services.) In their revised scam, the individuals were promoting coin collections to eager investors who were, unfortunately, mostly just elderly victims who succumbed to slick advertising and sales tactics. In their New York scheme, instead of selling "high-grade" coins, the swindlers were soaking used coins in a chemical solution to clean them up and then passing them off as coins of a higher value than they truly represented. (I could not find any classification of what a "high-grade" coin is, but the coin rating system is available on the Numismatic Guaranty Corporation's website at www.ngccoin.com. At the time of this writing, the NGC was listed as the official grading service of the American Numismatic Association, whose website is www.money.org.)

Part substitution frauds include the swapping of a more expensive part for a less expensive part and, in keeping the price the same, the supplier gets to pocket more profit. The impacts of this are more serious when the more expensive product being swapped is also of a higher caliber or performance. Aviation-grade nuts and bolts, ammunition, and some medical products are examples where the swapping of a more expensive and higher-performing product with one that is less expensive and has a lower level of performance can result in serious consequences such as injury or death.

Quantity Changes

Not all products are ordered in exact quantities. Items such as small parts might be sold by weight and not by number. Due to scale calibration and the characteristics (weight, volume) of the part, there will be variances in the replenishment quantity. The individual parts are so small that slight variances really do not matter and will tend to even out over time. Raw materials such as metal bars or spools of metal strips used in stamping processes will have similar variances between the ordered and replenished quantity.

These are somewhat natural or at least expected variances that both parties – the customer and the supplier – are aware of and adjust for accordingly upon receipt of the goods and in the amount of the invoice. Even if they are part of the nature of the product or process, these variances are not representative of fraudulent activity on the whole. Deliberate quantity changes (increases or decreases) on the purchase order are probably related to fraudulent activity. Falsifying the yield in order to deliberately steal product would be fraud.

An increase in the purchase order quantity (beyond what is actually required) may be the actions of someone who is looking to steal the overage amount. Depending on what is being ordered and whether the increases are small enough, they may go unnoticed for quite a while. Small parts might have higher accidental loss rates while sitting in inventory because of poor storage which results in spillage. (I have been in warehouses where small parts were littered around the floor and the people in charge wondered why their inventory counts were not as accurate as they would like them to be.) In some manufacturing environments, bulk quantities of small parts are moved to the various operational stations without much concern over exact quantities used versus those wasted or missing. As a result of all of this mishandling, who would really notice if someone was stealing small quantities of small parts? It would be wrong to equate small parts with cheap parts: Some of these small items are very expensive relative to their size. Unless audits attempted to reconcile purchases versus use, it is not likely that this pilferage fraud would be caught.

Fraudulently decreasing the quantity on a purchase order could be perpetrated with the same results in mind: To steal the excess quantity. However, in this scenario the purchase order quantity would be decreased in the organization's enterprise software after the order was already placed, with the excess quantity received – the difference between the original order quantity and the new purchase order quantity – ending up in the pockets of the fraudster. The invoice from the supplier would reflect the original purchase order quantity. Again, if the quantity differences were small enough, inventory control lax enough, and any audit or reconciliation procedures loose enough, this fraud could go on for quite some time.

Price Changes

As reported in the March 13, 2010 edition of the *South Florida Sun Sentinel*, the Taxi and Limousine Commission of New York City determined that approximately 36,000 cab drivers – representing 75 percent of the total licensed cabbies – defrauded 1.8 million riders of $8.3 million during a 26-month period. (A "rider" was someone who took a cab ride and thus the same person could have been defrauded more than once if he or she took multiple cab rides from any of the guilty drivers during the audit period.) The drivers who perpetrated this fraud did so with the flick of a switch by setting their meters to the higher out-of-city rate for in-city trips. This equated to an average overcharge of $4.61 per cab ride.

While the individual overcharges per trip were relatively small, it is easy to see how this fraud grew into a significant amount. Similarly small overcharges on repeated purchase orders can result in a sizable take. But who would perpetrate this type of fraud? I can envision a scenario where a person in a purchasing role increases the price of goods on a purchase order and the supplier splits the ill-gotten gains after the invoice is paid. If the accounting department is unaware of any pre-negotiated contract pricing, it is simply going to pay the invoices. If the fraudster is in charge of invoice approval, the accounts payables staff would have little reason to question the invoice amount. Having the accounting department review supplier contracts and have review – if not control – over the setting of vendor pricing in the ERP system would be a good check-and-balance procedure. Repeated and unnecessary set-up, handling, or other types of services fees added to the invoice could be representative of fraud in charging for fictitious services.

The following is an interesting story I first saw reported in the March 31, 2010 edition of the *South Florida Sun Sentinel*. The US Food and Drug Administration (FDA) was investigating what was being called an economic fraud against consumers who purchased frozen seafood. Ice is applied to frozen seafood as a means of protecting quality during storage and distribution. However, the complaint was that the weight of the ice was being included in the weight of the food itself; thus, consumers were being misled with regard to the true weight of the food they were purchasing. The initial investigation conducted by weights and measures inspectors from 17 states found that the ice coating accounted for up to 40 percent of the total product weight. Including the ice in the product cost effectively increased the price-per-pound being charged to the consumer.

This story provides a twist on product frauds and highlights the difference between product weight (the net weight on food and beverage products, for example) and shipping weight. Where applicable, customers must be aware of the weight of the products they are purchasing and be able to distinguish the product weight from the overall shipping weight. Suppliers have a right to recover overhead costs for packaging materials in their product pricing – they are valid business expenses just like rent, utilities, and personnel, and must be factored into the overall price. However, consumers must be informed of the actual product net weight that truly reflects the quantity being purchased without regard to packaging, wrapping, fill material, and in this case frozen water.

5

Distribution, Shipping, and Receiving Frauds

In many smaller organizations the picking and packing functions are performed by the same individual. Ideally the person who picks is not the person who packs, and this separation of responsibilities allows for the packer to audit the picker's performance. Tough economic times tend to drive organizations to reduce head-count by combining functions such as picking and packing; however, this creates more of an opportunity for fraud to be perpetrated.

Over-Picking/Over-Packing

Picking more than a customer order calls for is likely to be a simple case of theft. The pick list generated from the sales order listed a quantity of X and a higher quantity of Y was picked, the overage amount going into the proverbial pockets of the picker. Items can be hidden in trash bins, boxes of waste, and lunch pails, which all act as transport devices to get finished goods or raw materials out of a building. In organizations with loose inventory control, where cycle counts and physical inventories are rarely or poorly performed, this loss would probably go unnoticed, at least for a while. An over-pick is probably perpetrated by the person picking and does not involve collusion with anyone else.

Purposefully packing more than what an order calls for is also theft. The packer would probably need the picker's assistance if these two functions are separated. The overage quantity is shipped to the customer, who – in likely collusion with the packer – provides a kickback split to the packer (who would then have to share some of the illicit gain with the picker if the functions are separated). A person at the customer site who is likely to be involved in this fraud is someone in the receiving function who can intercept the overage quantity before it is noticed or checked into inventory or quality assurance.

Under-Picking/Under-Packing

If it is not possible to pick more than the order calls for, then it might be necessary to simply steal from the original order quantity and ship less than the amount ordered by the customer. If the customer is used to imperfect performance from the supplier, this fraud will be easier to get away with. If the product has a "natural" over/under quantity variation, then the customer would probably be unaware that anything untoward has occurred. A separation between the receiving and accounts payables processes restricts the accounting department from knowing that the supplier invoiced for a greater quantity than was shipped and received. For industries where supplier performance is closely monitored and measured, and where the data flows through the supply chain software application functions for cross-checking, even small thefts from an original order quantity can be caught and could result in financial penalties for non-compliance, having a negative impact on the supplier's scorecard rating, possibly resulting in lost sales or a lost customer.

Wrong Pick/Wrong Pack

Picking and packing the wrong quantities of an ordered product are not the only types of frauds relevant to these functions. Deliberately picking and packing the wrong product can have devastating effects on the supplier–customer relationship.

> A client of mine manufactured its own brand of product as well as private-label products for several US retailers. An employee in the distribution center – very unhappy with his job, his co-workers, and his supervisor – decided to exact revenge on the organization by deliberately swapping some of the retailer-specific private-label products between shipments to two different retailers. As such, Retailer A received some product labeled for Retailer B and vice versa. Neither retailer knew that my client was manufacturing product for other retailers, nor did that fact have to be revealed; many manufacturers (on-shore and off-shore) create private-label products for competing companies. The employee who caused this disruption most certainly was let go, but the damage was already done; my client's relationship with two key customers was placed in serious jeopardy. This incident could have been prevented if personnel and procedures had been put into place that separated responsibilities and acted as a check and balance to ensure that the right product was sent to the right customer.

Incorrect Pick/Incorrect Pack

Deliberately incorrectly picking or packing an order is an indication of one type of return fraud. In this scenario:

1. The customer orders Product A, but is shipped Product B.

2. The customer returns Product B, probably not having opened the package or having used the product.

3. The fraudster at the supplier takes in the returned Product B and classifies it as damaged or otherwise dysfunctional. Product B is not placed back into the finished goods inventory and is given an incorrect/false status of scrapped.

4. The customer is shipped Product A as originally ordered.

5. The fraudster steals Product B since it is supposedly damaged and cannot be restocked and resold.

Collusion with the customer may or may not have occurred; this fraud could be perpetrated either way. If it is done in collusion with the customer, care must be taken that repeated returns from a single customer do not raise any red flags in the supplier's order processing or fulfillment software. If the supposedly damaged products are submitted for claims by the organization shipping against the carrier the extra or excessive claims may be a red flag that fraud is possibly being committed.

Law of Commercial Transactions

In 1892 – at the recommendation of the American Bar Association – the National Conference of Commissioners on Uniform State Laws (NCCUSL) (www.nccusl.org) was established in the US. It is composed of legal practitioners (e.g. lawyers and judges) from each state who are appointed by each state's government. In the US, individual states are granted broad legislative authority over what happens within the confines of their own borders so long as state laws do not conflict with federal laws. The purpose of the NCCUSL is to draft legislation that, within the boundaries of the US federal government laws, permits intra-state consistency but also reflects each state's unique jurisdiction over what happens within its borders.

The Uniform Commercial Code (UCC) is one of the acts the NCCUSL is responsible for drafting and focuses on commercial transactions. Cornell University Law School has an excellent website dedicated to the various articles of the UCC (see www.law.cornell.edu/ucc). Article 1 contains definitions of terms used during the transacting of commercial commerce.

The UCC defines the bill of lading as "a document evidencing the receipt of goods for shipment issued by a person engaged in the business of transporting or forwarding goods." The shipper prepares the bill of lading for receipt by the buyer. This is different from a manifest, which lists all the bills of lading or waybills in the shipment and is created by the carrier (e.g. motor freight company) or an agent of the carrier.

Using the bill of lading guideline document provided by the US retail industry's trade association, Voluntary Interindustry Commerce Solutions (www.vics.org), we can see that the bill of lading contains specific information about the shipment, including from where, to where, by whom (the carrier), what is being shipped, whether the shipment contains hazardous materials, weight and volume, shipment method (carton and/or pallet along with counts of each), and other shipment-related information like the trailer and door seal identifiers.

The National Motor Freight Classification (NMFC) is a code structure established by the National Motor Freight Traffic Association (www.nmfta.org), which applies to intra-state, inter-state and foreign commerce. The purpose of the NMFC code is to group products together by the following four characteristics: density, stowability, handling, and liability. As stated on the NMFTA's website, the goal of the NMFC's 18 classifications is to provide "both carriers and shippers with a standard by which to begin negotiations and greatly simplifies the comparative evaluation of the many thousands of products moving in today's competitive marketplace." Each of the classifications represents the minimum requirements to ensure the safe transport of the goods. The transport of hazardous liquids requires more packaging and handling considerations compared with, say, the transport of feather pillows, and carriers are likely to adjust their rates accordingly based on what their customers are shipping. Similarly, the transport of refrigerated or frozen products (especially food) requires special transport capabilities and holding facilities, and one would expect to pay more for such services.

The pick list, the pack list, and the bill of lading are all documents primarily created in the distribution center. Ideally, the pick list is generated from the sales order, the pack list is created based on the inventory picked (with the possibility that an order cannot be 100 percent fulfilled), and the bill of lading is created from the packed shipment. Not all organizations have enterprise systems that provide this level of integrated control and, as such, these gaps can be exploited. For example, in many smaller to medium-size organizations, the pick list and the pack list are the same document and manual entries (that is, "handwriting") are used to update the actual quantities picked and packed.

Pick List/Pack List Document Frauds

In the pick and pack frauds previously discussed, the fraudster simply perpetrated the crime by relying on poor internal controls to avoid at least immediate discovery. There was no discussion of the fraudster altering any data to commit the fraud – it was simply done. One advantage of manipulating the data on the pick list and/or pack list is that a person looking to perpetrate fraud could essentially get someone else to do his or her dirty work, creating distance between the actual fraudster and the illicit activity. The guilty party would have to have oversight of the activities that are now performed by subordinates. The same pick and pack frauds could be carried out as before, but now there is a fall person to blame in the subordinate; the fraudster can deflect – at least for a while – the chances of being identified as the true culprit. It would take an audit of pick lists and pack lists by someone of a higher authority than our slimy supervisor or manipulative manager before the truth would be uncovered. A clever fraudster in this scenario would look to start and stop these frauds and alternate the victims among all the different subordinates under control to avoid any kind of pattern that could be a giveaway to the illicit activities being undertaken. The ability to manipulate the necessary data relies on our fraudster having the proper software application access; the ability to carry out this fraud for an extended period of time is enabled by the lack of audits and software transaction logging capabilities (which are either non-existent or not activated).

Bill of Lading Document Frauds

The discussion of frauds relating to the bill of lading introduces us to another possible participant in supply chain frauds: The transportation company, whether by land, by sea or by air. Unless the supplier or customer uses its own fleet, the transportation company will be independent of either party even if it is under contract with one or the other.

Because the NMFC code on the bill of lading is used to establish transportation rates by the carrier, falsification of this information could result in artificially higher or lower rates. A shipping manager could falsely claim his or her organization's products under an inappropriate classification code in order to obtain lower shipping rates, making him or her look like a real cost-saving hero to his or her management. Unfortunately the transportation company could be unaware of the true nature of the products it is carrying and inadvertent mishandling of the shipments might lead to serious consequences and regulatory fines. If an unscrupulous shipping manager can get away with it, falsely identifying goods at an NMFC code of a higher rate would increase the billings to the transportation company, which would likely kickback some of the ill-gotten gains to the inside fraudster. Without independent audits, who at the supplier's company would consider that they were being defrauded in such a manner? Falsification of weight and volume information can result in higher-than-normal billing, with kickbacks returned from the transportation company to the fraudster at the organization shipping the goods.

Shipping Frauds

Once goods leave an organization's shipping dock, they are in the trusted hands of the transportation company. Did all the cartons get on the truck or were one or two left behind? Was a carton counted twice during the staging or loading process? If so, was there an over-count or an even-count; if the latter, where is the missing carton? Things can go missing in transit. Accidents happen, but so does fraud. How? It is simple theft. Someone at the transportation company just walks away with a carton or removes an item from the middle of a pallet. It is easier to steal something from the middle of a pallet if the pallet is not shrinkwrapped, though it would not be that difficult to rewrap a pallet after such a theft. The observant fraudster employed by a motor freight company would be a very good frontperson to be on the lookout for shippers with lax procedures and hopefully high-value goods. It would be easy for the truck driver to steal some of the goods in transit. In collusion with the transportation company's distribution center personnel, goods can also be stolen during the cross-dock movement from one trailer to another. This supports the need for shippers to have strong procedures in place for shipment staging and loading and for those procedures to have verification steps to ensure 100 percent accuracy in carton and pallet counts. This information might be required as evidence in the event of an insurance claim against lost product or negative marks by a customer against a supplier scorecard. Because transportation companies are placed in the middle of trusted supplier–customer relationships, their own internal controls (procedures, security, and technology) must be of the highest caliber.

The blame game between the supplier and customer gets complicated with the carrier in between. The customer can blame the supplier for short-shipment or damages, with the supplier exclaiming it was not its fault but that of the shipping company that the customer has contracted. When the customer levels financial penalties for non-compliance against the supplier and the supplier is not at fault, the supplier is justified in shouting "fraud." More on the topic of supply chain vendor compliance will be presented later in this book.

Receiving Frauds

In a disorganized company, receiving frauds might just be the easiest frauds to perpetrate. Goods are received and some are immediately stolen. The receiving report is falsified to show complete fulfillment. To aid in covering up this fraud with the appearance that overall inventory levels are within normal parameters, other receipts for the same products – though possibly from different suppliers – might be falsified to show a higher amount than what was physically received. Without an audit or verification to purchase orders and supplier invoices, and without routine cycle counting and thorough physical inventories, this fraud could be perpetrated for a relatively long time. However, internal theft is not the only type of receiving fraud. A supplier that is knowingly looking to defraud its customer might deliberately short-ship but provide a kickback to someone at the customer's facility to falsify the receiving report.

6

Quality Assurance Frauds

It is one thing for a supplier to attempt to defraud a customer – either in collusion or independently – by shorting order quantities or faking returns, but fraud takes on a different nature when it is related to quality assurance. The fact is that compromising on quality can result in injury or death of the consumer or user of a product. While it may be tempting – or perhaps more traditional – to initially heap sole blame on external suppliers, quality assurance fraud can be perpetrated internally by employees of the organization.

Inside–Outside Collusion

Quality assurance fraud as perpetrated by a supplier is probably obvious: A supplier either cannot or is not willing to manufacture to the required specifications and provides bribes or kickbacks to a key employee or two at the customer organization to look the other way. The supplier cannot afford the new mold or the more expensive first-quality materials necessary due to its own economic problems, which may be the result of bad business decisions that could have been exasperated by personal vices or family pressures. The supplier may be outsourcing all or part of its manufacturing processes and may be lax in its own quality enforcement.

The supplier would have to reach into its customer's organization and then identify and change the behavior of at least one person in a quality assurance role. Tests of the supplier's inbound product would have to be falsified to cover up for the quality deficiencies. This can be accomplished by either not running certain tests or fraudulently changing the test results. Tests that are performed manually are easier to avoid and record false results against. However, in all probability it would not take too long before products fail and the cause is back-tracked to fake test logs, if manual record logs even exist. Automated tests should be supported by software that record the identification of the tester, the date/time of the test, and the test results against some type of tracking identifier such as the purchase order from which we know the supplier. Automated test results should be harder to falsify and easier to determine whether they were even performed or not.

Inside Pressures

Providing assurance at the right level of quality takes time and resources. It does little or no good to conduct validation if the tests are insufficient in scope, prone to error, or the sampling sizes are too small. The quality assurance area should not be viewed as a bottleneck to operational performance but as a necessity for operational excellence. It should not take that number of failures before a customer finds another source of whatever a quality-deficient supplier is providing. Suppliers can gain competitive advantages by implementing quality assurance programs and ensuring close supply chain relationships based on their performance. Especially for commodity suppliers, these aspects may be what separate them from their competitors and allow the supplier to retain current customers and acquire new accounts. Electronic components, pharmaceuticals, apparel, footwear, consumer electronics, and food and beverages are examples of industries where there are generally ample supplier choices.

Quality assurance programs that are viewed as operational roadblocks or are unresourced to the point of being bottleneck processes will cause stress to other functional areas such as manufacturing and distribution (order fulfillment). If manufacturing cannot get its raw materials on demand or order fulfillment cannot get its finished goods when it needs them, it could result in overtime and delays. The manufacturing, distribution, and accounting departments may put pressure on quality assurance to speed things up because their delays are causing increased costs, angry customers, and supplier demands for payment, the latter situation occurring if the organization does not pay the supplier invoice until all quality testing has been performed. If the quality assurance manager is at a hierarchical disadvantage on the organization chart, he or she might have little leverage and be forced to yield to the pressure from within the organization.

Such internal pressures can force compromises in the quality assurance process. The quality manager might become fearful of losing his or her job, perhaps even having been threatened with dismissal unless "improvements" are made. Unfortunately, if the organization fails to resource the quality department adequately, there is probably nothing the manager can do to rectify the situation. Remember that fraud is defined as a breach of confidence. Employees and their employers must have confidence in each other with regard to job roles, responsibilities, and the work environment.

Customers have an expectation that the products they purchase are of a minimum acceptable standard based on the industry and type of product, and that these products will not cause damage or injury when used as they are supposed to be. This applies to the ingredients used to make, for example, pharmaceuticals, food, and beverages, as well as the finished products themselves. Within the organization, manufacturing and distribution are customers to quality assurance and should have confidence in the ingredients and finished goods they are receiving into their respective areas. The organization that fails to properly resource its quality assurance department – be it people, testing equipment, software, or facilities – and allows it to be a problem area is simply guilty of breaching confidence between itself and its employees and customers alike. By purposefully placing others at risk due to known compromises or should-have-known-better actions (or perhaps failures to act would be more accurate), an organization can be guilty of fraudulent behavior.

Quality Fraud in Disposals

Hazardous waste from manufacturing operations is common in industries such as metal stamping, power generation, textiles, and food and beverage production. A byproduct of the orange juice manufacturing process is an extract of orange oil called d-limonene that is so strong that is it sold as an industrial solvent and cleaning product. Selling off such byproducts is preferable – and hopefully more profitable – than simply dumping them in fields or streams and polluting water supplies. Similarly, the output from product reclamation operations might produce a waste byproduct as in the breakdown of consumer electronics: Some of the components contain hazardous metals that cannot simply be buried in the ground, even at waste dumps.

The incorrect disposal of a toxic waste or hazardous byproduct – especially if there are regulatory rules in place – is fraudulent behavior. Just as quality assurance tests are critical to ensure what comes into or is produced by an organization is up to par, quality assurance tests are needed to ensure what an organization disposes of is filtered, reduced, diluted, and/or otherwise contained as to no longer be a health hazard. Employees might be the subject of internal organization pressure to falsify quality tests in order to pass regulatory requirements. Occasionally there are reported cases of syringe reuse in health care facilities. This may merely be the result of poor procedures and a lack of common sense, or it could be the result of cost-cutting directives by senior management. The reuse of disposable medical equipment is fraud, as would be the failure to properly sterilize certain medical instruments in between use.

7

Inventory Asset and Fixed Asset Frauds

When an organization acquires something tangible, it is an asset. Inventory assets are used to make a finished good, support a service, or to be resold. Raw material plastics and metals are examples of inventory assets. In a hospital, bandages, tape, syringes, crutches, and pharmaceuticals are all inventoried to support the medical services provided by the hospital to its patients. A retail store inventories finished goods for resale to consumers. Inventory assets are not intended to stay with the organization for a long period of time; ownership is meant to be relatively brief. Fixed assets such as manufacturing machines, furniture, computers, medical equipment, hospital beds, etc. are typically owned by the organization for years until the end of their useful lives.

Both inventory assets and fixed assets share the common characteristics of location, original cost, present value (subject to depreciation), usable life, and movement. Inventory assets are typically located in a warehouse, storeroom, or retail store down an aisle, on a shelf, and maybe in a bin. Fixed assets are usually located in a building, on a floor, in a room (classroom, hospital room, office, etc.) or cubicle, or in a fixed position in a manufacturing environment.

One of the most – if not *the* most – common asset frauds is theft. It is much easier to steal something when there is little or no control over acquisition, location, and movement. Expensive inventory assets such as precious metals, jewelry, and electronics, and controlled substances such as pharmaceuticals should be stored in locked rooms or metal cages with restricted access. Fixed assets should be secured when possible, such as cable-locking a personal computer or printer to a desk. The relatively inexpensive cost of personal computers makes this statement seem almost ridiculous, but even slightly used computers have some value. It is worth bearing in mind that the thief may be more interested in the information on the hard drive than the physical computer itself. Therefore, securing the computer case to a desk and preventing the computer case from being easily opened are ways to protect both levels of this particular fixed asset.

Aside from securing assets and restricting their movement, cycle counts and full physical inventories are traditional methods of ensuring that everything is accounted for and they remain necessary control activities. Serialized assets should be tracked at this level of detail. Food, beverages, and pharmaceuticals are typically identified with lot or batch control identifiers and expiry dates, and this information is useful not just for recall purposes but also extra internal controls. Electronics such as personal computers, printers, medical equipment, and cellular telephones are all identified with unique serial numbers.

When a fixed asset is assigned to an employee, there should be record of the asset and serial number in the employee's personnel file, and this information should be updated whenever the asset changes. Upon replacement for upgrade or damage repair, or due to dismissal or voluntary resignation, the employee must be required to account for and return the current fixed assets that he or she was most recently assigned. The issuance of a laptop computer or cell phone means that the laptop's hard drive serial number and the cell phone's memory card serial identifier should have been recorded upon fixed asset assignment to ensure that the employee has not removed or swapped these critical components when the asset is returned. Components that hold information, such as cellular telephone SIM cards and computer laptops, may be swapped to facilitate the theft of the information contained on them and not just for the component itself.

Employees should be held responsible for fixed asset movements as part of their performance. Organizations must create policies and practical procedures to ensure that employees are able to conveniently record the movement of a fixed asset. For institutions such as universities or hospitals that receive public funding, this helps to protect the tax dollars at work. People tend to get very irate when they read about stolen tax dollars, whether in the form of cash or assets. Fixed asset management also means fewer headaches and time wasted searching for something, whether it is audio-visual or laboratory equipment at an educational institution's campus or medical equipment in a health care facility.

Organizations that rely on perishable inventory assets have a greater level of responsibility in ensuring that the asset is stored within the necessary environmental parameters and is used no later than the expiry date. The failure by an organization in either scenario could be labeled as fraudulent behavior. Spoiled edible ingredients could result in illness or death and old pharmaceuticals may lose their efficacy.

8

Manufacturing Frauds

We have examined frauds that affect manufacturing, such as those involved with lapses in effective quality assurance programs or compromises in purchasing raw materials and components that meet minimum requirements. Now we will review manufacturing frauds that involve what is happening on the shop floor and can be perpetrated distinctly and without fraud occurring in other areas.

To recap, asset misuse or abuse is fraud. Recall the maintenance person who was not doing his job, the result being that machine control unit air filters had never been changed. An employee who misstates that he or she has performed a task – especially one that is at the core of his or her job role – is providing a false (fraudulent) statement. Granted, there are assumptions that the employee was qualified for his or her position, the employee's role responsibilities were clearly defined, and the employee was provided sufficient training to be successful in completing the assigned tasks. All things being equal – and in this case they were – the maintenance employee perpetrated asset abuse. Abuse of manufacturing equipment that causes less-than-first-quality product to be produced is fraud. Intentional overuse of an asset or using an asset for purposes other than those for which it was designed is fraud.

There are frauds that can happen on the manufacturing floor that do not involve direct asset misuse or abuse. In certain industries, such as metal stamping or the melting of raw metal to create molded products, the waste output – the fragments of precious metal or the leftover not used – has value. Theft of coils of cable from work areas, catalytic converters from automobiles, and components from ground-based cable junction boxes are not uncommon; the metal is taken to a scrap yard or recycle center and sold. These thefts are perpetrated under cover of darkness or sometimes in daylight as brazen acts. Speed is important, as is concealment of goods if they are being removed from a facility. Burying goods in trash containers in order to get them out of the building is one method of concealment. Lax controls and security procedures enable these thefts to occur. A person working in a manufacturing environment where metal bricks or coils are used could not only abscond with the normal waste but in some instances could cause excess waste to be produced, increasing the monetary take on the ill-gotten materials. The organization would then have a lower yield on materials, thus increasing costs. This is one reason why manufacturing companies that use valuable raw materials must have a detailed bill of materials by operation and be able to account for – and accumulate – the scrap from operations. The organization has the right to recoup its own raw material investment by selling off or recycling the leftover metals from its manufacturing operations.

Conversely, theft could occur during the acquisition of raw materials for manufacturing from inventory but before the operation begins. For example, bear in mind that pharmaceuticals – especially those in pill or capsule form – are manufactured in batches of hundreds or thousands. Exact measurements of the ingredient and no waste in the process would be required to ensure complete usage of the ingredients. This also relies on proper material handling to ensure that no spillage of ingredients occurs due to improper handling, leaks, or overflows. Extreme control measures would be required to ensure that no extra portion of an ingredient was acquired from inventory or siphoned off from the standard measurement before manufacturing began. If the story of taxi drivers who defrauded riders an average of less than $5 per ride shows us anything, it is that some fraudsters are patient and willing to sacrifice initial large gains for smaller gains over a longer period of time. In some manufacturing operations bulk inventory is moved to the specific manufacturing operation and the unused portion is returned to inventory. This represents another opportunity point for fraud to occur.

Even with extremely accurate weights and measures of ingredients or raw materials coupled with well-maintained machines, the theft of finished goods off the production line can occur. It might be tough to grab and conceal a can of food or a bottle of a liquid directly off the production line, but pills or small components could present greater opportunities for the occasional swipe. The less exposure to human hands, the smaller the chances of theft occurring; any exposed conveyor lines should be covered not just for safety but also to block any attempts to pilfer product.

Maintenance is more than merely making sure that a manufacturing machine runs well and that it counts accurately too. It is important to ensure that the machine's counter cannot be tweaked to give a false reading. If the machine says 100 pills went into a bottle, is that true? Is the bottle weighed before or after sealing or capping? Can it be known whether 100 and not 99 pills went into a bottle? A batch run that reduced the pill count per bottle by one would leave leftover pills that were not bottled and therefore could be stolen.

In order to combat manufacturing frauds, the organization needs to know how much of each input is required to achieve the desired output. More formally this is known as a Bill of Materials. For each operational process (defined by the Bill of Operations), the amount of input (raw materials or components) is recorded. Just as in a recipe, the amount of each ingredient determines how much will be produced. Depending on the quantity of output needed, the amount of each input or ingredient needed should be calculated based on a pre-determined factor from a basis unit (e.g. how much it takes to make one of something) that represents the smallest output quantity. Maintaining uniformity and control in how the quantity multiplier is applied ensures that the amount of ingredients required is not falsified in order to allow a fraudster to attempt to withdraw more of any one ingredient.

The Bill of Labor is a record of how much human effort is required at each operational process. Again, knowing what it takes to make a base quantity, it is possible to calculate the amount of effort to make multiples of the base quantity. By keeping records of the labor effort expected and comparing these to the output produced and payroll records, an organization can create a system of checks and balances to audit for fraud relating to payroll and employee time and attendance.

The inability to control materials and labor can result in fraud against the organization and can adversely affect the supply chain, resulting in the organization having to acquire excess ingredients (with the potential of excess freight charges) and distribute excess payroll in the form of pay for less work than the employee is performing or for more employees than the organization requires due to reduced employee productivity. Further, if some ingredients are controlled substances or hazardous materials, the organization may find itself breaching regulatory requirements for its lack of controls.

9

Invoice and Sales Commission Frauds

Purchase order frauds can result in fictitious invoices from real or fake suppliers. Submitting an invoice from a real or fake supplier without supporting documentation – notably the purchase order – leaves much doubt with regard to whether the invoice is valid. This gap is perfect for exploitation in organizations that do not have comprehensive enterprise software or are prone to doing business "the old fashioned way," such as via oral communication by telephone with nothing in writing, whether in electronic or paper format.

To give an example, a supplier could infiltrate an organization's accounting department and conspire with an inside person under similar circumstances to other inside–outside collaborations already discussed. The purpose would be to increase the supplier's payment amount beyond the invoice submitted, with the difference (the ill-gotten gains) split between the supplier and the accounting person. An audit that reconciled the documentation (purchase order and invoice) to the check register would be required to detect such a fraud.

A supplier who sends an invoice for the entire order but knowingly only shipped part of the order might be guilty of perpetrating fraud, although this depends on the terms of the relationship between the buyer and supplier. It may be fraud to collect more money in the short term and then ship the remaining product. It would more probably be fraud if the back-order was never fulfilled and no refund or credit was given to the customer. A supplier who deliberately does not apply credits to invoices may be looking to defraud its customers, only noting the credits when they are detected.

The customer's payment of a supplier invoice can be the trigger for the supplier's payment of sales commissions to their employed or independent sales representatives – once the invoice is paid, the deal is done. Frauds related to the payment of sales commissions can include payments based on fictitious orders from either real or fake customers, payments on booked and not actual sales, and payments on sales that were returned later.

Sales commission fraud does not depend on collusion with the customer, but such collusion would make it a lot easier to perpetrate as far as the scheme itself goes. Without customer cooperation, a sales representative would have to falsify a sales order, which may not be that difficult to do. However, when the order ships and is summarily rejected by the customer, the sales representative's fraud would be discovered. Even if the goods make it to the customer and sit around while an investigation is performed, I hope that the customer would not pay the invoice. Paying sales commissions after the customer invoice is paid is one method of ensuring the authenticity of the original sales order.

A means of perpetrating sales commission fraud is through order stuffing, which was covered earlier. The customer takes in more than he or she needs over time and receives a kickback for doing so. The problem with order stuffing is that at some point no new customer orders will be generated and the excess inventory held by the customer will have to be depleted before any new orders will be forthcoming. With no new orders, there will be no new sales commissions, so order stuffing is a relatively short-term solution to gain more immediate funds. However, if the customer organization returned the excess goods, it would be relieved of the extra inventory and get money back. It is true that the customer organization would not have use of that money for a while, but the fraud perpetrators would not be worried about that. If sales commissions are not reimbursed when there is a customer return, a sales representative in collusion with a customer could be successful in committing this type of fraud.

10

Return Frauds

Fraudulent returns are somewhat different from return frauds. A fraudulent return could be used to perpetrate inventory theft, the stealing of funds due to the submission of fake invoices, or sales commission fraud, and generally involves the product "as is." Return frauds are those that rely on an alteration made to disguise the fact that the original item is not truly the one being returned. The following are examples of different types of return frauds.

Removed Components

As its name suggests, this fraud involves the removal of components from an item before the item is returned. Electronic gadgets are an easy target for this fraud with their memory modules, mother boards, audio-video cards, and other components.

Swapped Components

The problem with simply removing components is that it is a little obvious upon inspection. Components might be swapped with non-working replacements to mask the theft of the original's good components. An equivalent component that is either new or used but that does not meet the manufacturer's product specifications can also be substituted for an original component. The product is likely to be returned as dysfunctional upon receipt or damaged during use and thus would not be expected to work upon return.

How about health and weight loss products in capsule form? Some of the advertisements allow for full refunds upon return of the unused portion. It might take some effort, but the original pills could be swapped for duplicates containing something similar in color and texture, such as flour or baking soda.

The fraudster could then retain the original product, sending back the fakes. The fraud would be discovered only if the company tested a sample of the returned product before issuing a refund. Fraud doesn't always have to make perfect sense: The fraudster may perpetrate an act merely for the satisfaction of doing so rather than for any great financial gain.

Missing Product

If removed component and swapped component return frauds were not convincing reasons for establishing an inspection process before a customer refund is issued, surely having something completely different returned from the original product is reason enough. The supplier should make sure that the customer has returned the product that he or she purchased and not a rock, a bag of shredded rags, old electronic equipment, or anything else that has a similar weight and feel inside the product box.

Out of Warranty Product

Unless an exemption is made on a customer-by-customer basis, products no longer under warranty are not eligible for full refund upon return. This is an excellent reason why certain products should be assigned a unique serial number or alphanumeric identifier that is tied to the order fulfillment process and is made available to customer services. Companies with warranted products may be required to hold specific cash reserves to cover anticipated losses, performing similar calculations and forecasts as are done with insurance-industry actuarial tables. High incidences of warranty fraud can skew the forecasts and can force the organization to hold more cash in reserve than would normally be necessary, because more claims are being paid than should be and were forecast based on the product's warranted life. Inasmuch as a company may want to provide good customer service, it must also protect itself from unscrupulous consumers who would seek to defraud a company by making a claim against an out-of-warranty product.

Used for New

Let us recall the fraud story of old coins essentially being washed and cleaned up to be sold as higher-value items then they truly were. Similarly, one type of return fraud would be to substitute all or part of an item with the same component of used instead of new condition. Electronic gadgets might have their new components – hard drives, memory cards, video cards, etc. – swapped with used components. The returned item could be made to be sufficiently functional with the used components to pass a cursory inspection. Without at least some level of quality inspection, a comparable product could be shipped back instead of the higher-priced new product the consumer originally ordered. In this latter case the organization might have to tread a little carefully in terms of accusing the consumer outright of fraud; the consumer could claim that it was an accident that the comparable product got into the box where the new product should have been.

More difficult to spot might be the returned item that was only used once or twice, perhaps even being returned after having been professionally cleaned in the case of apparel and accessories such as suits, jackets, dresses, and purses. Forcing the consumer to remove outside tags is one way to try and catch these people in the act. However, this relies 100 percent on the items being labeled with a warning to the consumer that the removal of the tags disqualifies the item from being returned. And while I am sure this is what the retail industry would desire, I understand that in the world of consumerism it is impractical. Retailers can protect themselves best by limiting consumer returns based on the number of returns and within a specific number of days from the ship or receipt date.

Returns – involving reverse logistics – is a supply chain in and of itself. If properly controlled, a returns supply chain can highlight product problems and catch consumers who seek to take advantage of an organization's gaps in operations and software applications for return authorizations. Analytics from the returns supply chain should be used to determine if there is a problem that exists in the inbound or outbound supply chain that could result in product deficiencies, such as if a product does not perform to the standard that it should or performs to its standard but not for the anticipated life of the product. The analytics should be used to trace back to the suppliers of raw materials, components, and services, keeping in mind that the suppliers can be internal or external to the organization. The cause of the return could be defective materials or a defective process somewhere in the inbound or outbound supply chain.

11

Detection Do's and Don'ts

Detecting supply chain fraud or any other illicit activity requires sniffing out clues while traveling down what will probably be several different investigatory paths, finding evidence, and building a case. Evidence can be material or circumstantial, though certainly the more firm the evidence, the more solid the case against the fraud perpetrators. In the business environment, circumstantial evidence of fraud can trigger investigations and can lead to the detection of gaps and the introduction of better controls. Circumstantial evidence, when presented to an employee, may be enough to get an employee who was involved in the perpetration of fraud to agree to resign on his or her own; guilty parties may welcome the chance to avoid the involvement of law enforcement agencies if the organization decides to get them involved. Unfortunately, the act of letting go an employee known to have perpetrated fraud leaves the fraudster free to pursue employment elsewhere and possibly run his or her scams again, the defrauded organization passing along the problem instead of stopping it. Organizations may wish to involve law enforcement agencies and press charges against their dishonest employees, but too often (and unfortunately) they may lack the will to do so; the involvement of law enforcement agencies might result in the case being reported in the news, which could negatively affect the organization's reputation, customer or supplier relationships, and perhaps even stock price. Even where the burden of proof is lower in civil versus criminal litigation, the fact that fraud occurred in an organization may make the organization reluctant to pursue any legal action for fear of the litigation being reported in the news media. I believe this is why we do not hear or read about more supply chain frauds in the media unless they are related to the abuse of taxpayer monies or involve consumer injury or death.

For the few years that followed my career change, I still kept in touch with some of my former co-workers. One day I received a telephone call from an old colleague who informed me that several members of the information technology staff, one person in a purchasing role, and an executive assistant had all resigned suddenly at my former employer. Apparently they had all conspired to falsify purchase orders with regard to personal computers and related hardware components. The group was found to have stolen approximately $100,000 worth of computer parts, which were being used to support the outside computer hardware business run by several of the group members. I was informed that instead of the organization bringing in law enforcement agencies and filing criminal charges, it was strongly suggested that each person tender his or her resignation or face dismissal, and apparently each person submitted his or her own resignation shortly thereafter.

There is no guarantee that an employee found to have committed fraud will yield under pressure when presented with evidence – circumstantial or direct – and resign his or her job. It is possible that the employee knows how flimsy or firm the evidence is in reality and is therefore more able to push back against the organization. Experienced fraudsters or employees with detailed knowledge of an organization's software applications and business procedures may have the wherewithal to know how much factual or fictional evidence the organization was truly able to acquire and thus may not be so swayed or threatened to resign when confronted by the organization.

The burden of proof in civil litigation is the preponderance of evidence, essentially meaning that the weight of the evidence will determine the outcome of the trial. If the plaintiff's evidence carries enough weight, the defendant can be found guilty; if the evidence is not convincing enough, the plaintiff's case may be seen as too weak and the defendant will probably be presumed to be innocent. In criminal litigation there must be evidence beyond a reasonable doubt, which basically means that a reasonable person would be able to make a determination based upon the facts presented. Evidence in criminal litigation must pass a stricter, more definitive test. Still, even with a reduced burden of evidence, civil trials can make the news and thus make organizations wary of pursuing them and getting unfavorable media coverage.

Several days after I presented to a group of manufacturers on supply chain fraud, I received a telephone call from the symposium organizer. Following my presentation, one of the attendees went back to her company and informed the owner that she suspected a fraud.

The company owner began his own internal investigation, but unfortunately mishandled it to the extent that he immediately revealed to the suspected fraudster that his (illicit) activities were being closely scrutinized. Unsurprisingly, the rogue employee began covering his tracks and ending his operations. In the end the fraudster simply left the company. The owner still does not know how much was stolen: Somewhere between $60,000 and $250,000 is the best guess. Insufficient controls, a lack of proper record-keeping, too little oversight, and too much trust by the company owner in this employee were all likely contributors to the inability to better determine the extent of the fraud.

The company owner was reduced to tears over the revelation of fraud. The fraudster had worked for the company for many years and had perpetrated his fraud through both the company's good times and bad times. Fraudsters rarely care about the financial fortunes or failures of their employer as they commit their crimes; their priority is their own needs first and foremost. The owner believed he had treated this employee, like his others, very fairly, but that is subjective and beside the point. The fraudster saw the opportunity, had a need, and took advantage of the situation. By muddling the investigation, the fraudster was allowed to go free with the ability to perpetrate fraud at his next employer. The defrauded organization was left unable to pursue him through the law, which may also prevent it from filing any insurance claims to recoup all or part of the losses.

The Dangers of Do-It-Yourself Investigation

What contributed to the above investigation getting derailed and prevented the organization from involving law enforcement agencies and filing criminal charges such as theft is that the organization made the not uncommon mistake of performing its own investigation. Fraud investigations are not do-it-yourself projects to be undertaken by those without the proper skills. While both a parking enforcement officer and a homicide detective are each performing duties related to the investigations of illegal activities, there are considerable differences in the job requirements and qualifications. A financial auditor, network supervisor, or hardware technician may be the wrong people – at least individually and initially – to assign to a suspected case of supply chain fraud.

The failure of an organization to investigate supply chain fraud with due diligence and care can result in a failure to find the fraud or a mishandling whereby the case itself cannot be pursued to the necessary extent. An investigation of an employee suspected of perpetrating fraud might require a forensic analysis of the hard drives or memory cards of the PC workstation, laptop, or cell phone assigned to the employee. A failure to properly protect the hard drives during their removal from the computer can damage them or cause damage to the drive contents, such as that which could happen in close contact with a static electrical charge or exposure to excessive moisture or heat. A "true copy" of the hard drive should be made to protect the purity of the original source hard drive. Any forensic analysis should be performed against verifiable copies of the original hard drive, leaving one "true copy" always intact and unchanged. And it is important to remember that opening a file changes the file's date/time stamp from the last time it was accessed, e.g. by the suspected fraudster. A chain of evidence must be sufficiently established and very well documented, including who was in possession of what and when (date and time). Any gaps or lapses in procedure could be used by a defendant to throw doubt upon the veracity of evidence by the plaintiff organization.

Use Evidence Trails

Computer networks (the technology infrastructure) can tell us the date and time of a person who has signed in or out and which terminal or client computer he or she has used. Was it the person or someone who learned of another person's login credentials (identification and password)? Good security policy will inform users to lock or otherwise disable access to their workstations when they leave them to help ensure that no one else can access information using their identification credentials. Employees should be informed to protect their computer system and software access credentials (their user identification and password). Some application, operating system, and network control software can be configured to prompt users to change their passwords on a regular basis. Secure access doors record the date, time, and location of employees to enter (and maybe also exit) based on a badge swipe or other unique identification token.

It is worth recalling in the discussion of supply chain frauds the need for transaction and data maintenance logging in enterprise software applications so as to make it possible to catch certain activities. This functionality – again, when activated – provides a critical trail of evidence that is an important component of detecting supply chain fraud based on what a person is doing once he or she is inside the protected infrastructure. Organizations use product identification devices such as radio frequency identification tags, lot/batch identifiers, and serial numbers (the latter two most likely in barcode form) for track-and-trace capabilities through the supply chain, recording a unique identifier's location and date/time stamp as it passes through the supply chain. The ability to trace a product and the related transactions through the supply chain are parallel trails of evidence that can be used together to trace fraudulent activities.

During a consulting assignment, the sales manager brought me into a conversation with a sales representative who accused another sales representative of reassigning ownership of several customer accounts for the purposes of receiving commission on sales to those customers. The company's ERP system tracked the before and after values of data fields in transactions such as sales orders, purchase orders, and invoices, as well as against entities such as customers, items, and vendors. Thus, we were able to investigate the original ownership of the customer accounts and move toward determining whether or not they were fraudulently changed and assigned to the other sales representative. We found evidence of sales representatives who fraudulently assigned themselves to customer accounts; those sales representatives were soon dismissed after this was brought to the attention of the company executives.

Alerts from software applications are not the only fraud discovery sources on which organizations should rely. Let us recall the story of the children and families agency supervisor who knew not to withdraw payment amounts greater than the apparent $900 limit. Software transaction logging and controls – necessary and effective as they are – cannot stand alone as the only means of detecting fraudulent activity. In this situation the fraudster was aware that attempts to withdraw payments above the $900 limit would bring attention to her scheme by requiring her to receive approval for an amount greater than the $900 limit. Thus, she simply withdrew payment amounts less than the control limit.

In many cases an organization's employees, customers, or suppliers may report fraudulent activity that the internal audit or security department itself has failed to notice.

In its *2010 Report to the Nation*, the Association of Certified Fraud Examiners reports that tips "were the most common means of detection in every study since 2002, when we began tracking the data." In the *2010 Report*, the following statistics were presented with regard to the sources responsible for the initial detection of frauds:

- Tips: 40.2 percent.
- Management review: 15.4 percent.
- Internal audit: 13.9 percent.
- By accident: 8.3 percent.
- Account reconciliation: 6.1 percent.
- Document examination: 5.2 percent.
- External audit: 4.6 percent.

- Surveillance/monitoring: 2.6 percent.
- Notified by police: 1.0 percent.
- Confession: 1.0 percent.
- IT controls: 0.0 percent.

Further, employees were responsible for more tips (49.2 percent) than customers (17.8 percent) and vendors (12.1 percent), with an anonymous tip source representative of 13.4 percent of the tips. This highlights the need for the organization to really engage its employees, customers, and vendors in the fight against fraud.

Organizations that create integrated and respected relationships with their customers and suppliers can leverage this association to help each other combat fraudulent activities because all parties can be affected. There is also a chance that fraud between supply chain partners can be contained. Frauds caught early enough in the supply chain may be able to be dealt with internally or at least "less externally" than those that result in a full-blown crisis. It is more painful when an organization finds out about its fraudulent activities first through consumer groups, government regulators, or the media. The organization is blindsided to the news and is forced to react suddenly, perhaps even with damage control efforts if there are safety issues with a product it produced, was involved in producing, or distributed. Reacting to fraudulent behavior is much more expensive than proactively reducing the chances of it occurring in the first place.

Detection Without Disruption

Organizations need to combat fraudulent behavior proactively, which requires detecting it as early as possible. Controlled access to physical areas, security cameras, network and application security are all various means of reducing the possibility of fraud occurring. These are somewhat static, because a person either has access or does not. Business operations through a supply chain are fluid and dynamic, requiring a cross-analysis of information and transactions to determine whether activity is of a fraudulent nature. Detection frequency must be sufficient without becoming an obstacle, yet it must be continuous and pervasive throughout the supply chain. Depending on the industry and how an organization has developed its relationships with its suppliers, it may not be practical or necessary to open every box or sample every container, but it is necessary to check statistically significant amounts or quantities to allow confidence that an entire batch or shipment is complete and not tainted. Establishing quality control points through the entire supply chain removes the burden of complete quality inspection from one functional area and allows it to be shared across the enterprise. This is necessary because each step in a supply chain can add something (e.g. handling, an ingredient, a component) that changes or affects what is being produced. The goal is to increase the likelihood of catching fraud while decreasing the impact should fraud happen. The detection methodology should be somewhat stealthy so as to not reveal fully the tactics and controls used, making them more resilient in the face of potential fraudsters. Disruptive detection tactics run the risk of affecting operations negatively and thus skewing performance results, becoming more of a hindrance than a help. Detection tactics that are barriers to operational performance may force employees to work around them to maintain throughput and timely schedules, thus rendering the detection tactics ineffective and a contributor to circumventing systems and processes, resulting in even wider gaps where fraud can be perpetrated. For example, running too many analytical programs can tax computer system resources, while bombarding managers and supervisors with excessive statistical information is inefficient to the point of being counter-productive. Organizations must elevate the perception of detection to a level at which employees, customers, and suppliers have an awareness of anti-fraud oversight without feeling threatened by it and without being affected by it.

A good rule of thumb for supply chain operations is "trust but verify," a phrase attributed to President Ronald Reagan, although it is actually based on a Russian proverb. Reagan used this term during a speech with regard to a treaty reducing nuclear weapons between the US and Russia as Mikhail Gorbachev was standing next to him.

On October 14, 2008, Public Law 110-417, also known as the "Duncan Hunter National Defense Authorization Act for Fiscal Year 2009," was enacted by the US Congress. Section 254 of the Act is titled *Trusted Defense Systems*. The section states that: "The Secretary of Defense shall conduct an assessment of selected covered acquisition programs to identify vulnerabilities in the supply chain of each program's electronics and information processing systems that potentially compromise the level of trust in the systems." In other words, trust but verify.

Organizations can outsource manufacturing and other operations, but this does not translate to an outsourcing of responsibility. Regardless of whether an organization's supply chain is local or global, there must be procedures and systems in place to mutually ensure quality at each section of the supply chain.

The US government recognizes the need to trust other governments – even those once considered a fierce enemy – in the same way as the US military must have trust in relationships with its suppliers. Both entities recognize the need to verify those trusted relationships, and this is a good lesson for organizations charged with providing goods and services to consumers of all types.

The balance in deciding the appropriate level of detection is somewhere between how much is too much to the point of overwhelming and no longer being conducive – if not detrimental to – achieving the necessary purpose, and how little is sufficient to be adequate in covering a substantial amount of the problems. Some determining factors include the following:

- Industry: Food and pharmaceuticals demands a greater level and broader approach to fraud detection than, for example, certain commodity consumer products. This is not to say that every industry should not demand and drive towards 100 percent perfection. However, some industries are more tightly controlled via government regulations and have – by their nature – the potential to cause greater harm to the general public due to the failure to detect fraudulent activity within their supply chains.

- Employee education: Organizations should educate their employees as to why fraud detection and reduction are so important. Employees should be engaged in the fraud detection practices of the organization and be given the opportunity to anonymously blow the whistle on other employees suspected of fraudulent activity. As reported by the Broward Bulldog (www.browardbulldog.org) and carried by the *South Florida Sun Sentinel* newspaper on June 13, 2010, an anonymous whistleblower has alleged that several Broward County Transit employees were involved in the awarding of a $13.3 million no-bid contract involving unproven technology that was deliberately misidentified in transit commission records. Florida state law shields the identities of whistleblowers – whether employees or others – when a written complaint is filed, and Broward County has a "zero tolerance" policy toward whistleblower retaliation. Through these methods of protection, individuals can more freely report the possibility of fraudulent activity. In its *2010 Report*, the Association of Certified Fraud Examiners notes that in organizations with employee hotlines for reporting suspected fraudulent activities, 47 percent of frauds were detected, with tips versus only 34 percent of frauds in organizations without employee hotlines. Further, the report notes that organizations with employee hotlines were able to detect fraud seven months earlier than those organizations without an employee hotline.

- Method of detection: Visual inspection, chemical analysis, and weight comparison are various methods of detecting fraudulent activity in the movement of goods. Within each method there must be the appropriate investments made to use sufficient technology and equipment to be able to adequately perform the tests needed to determine fraud. The investment extends to ensuring that proper maintenance is performed on the testing equipment and that the maintenance records are subject to audit. Obvious as it is to state, weight comparison tests are less reliable when incorrectly calibrated scales are used. Employees involved in this process can become complacent due to the known failures of the equipment used in the detection process and the inability of employees to perform their tasks at the desired level of standard.

- Supplier relationships: Organizations that cultivate close relationships with key suppliers – including open discussions about inspection procedures and sharing data to help detect and reduce fraudulent behavior – should be able to adjust the number of inspections performed without severely compromising the quality of the inspection process. This can aid in focusing an organization's employees on completing even more thorough inspections in the same time period by removing the burden of too many inspections.

An organization's customers and suppliers will benefit as the organization shares its fraud-monitoring data with its supply chain partners. In such a collaborative relationship, partners may also agree to share their own fraud monitoring methodologies, though these can be considered proprietary information that an organization may opt not to reveal, so the sharing of information may be limited to just the results. Employees will be made aware of their fraud-catching performance or ability to spot suspicious activities through open communications from their respective management and as part of their employment review process. The employee's employment may depend on his or her ability to detect abnormal activity in the case of an audit or inspection job role. This is predicated on the organization deploying the necessary equipment and technology to enable the employee to detect fraudulent behavior. All things being equal, if suspicious activities continue to slip through under a particular employee's watch, it is unlikely that this employee is right for a job role that involves inspecting goods upon receipt from a supplier. (Conversely an employee who falsely accuses a supplier of fraud – perhaps to prove he or she is doing a good job of inspection – is committing an act of fraud and should be removed from the job role.) Suppliers will know that they are being watched as inspection audit information is shared with them by their customers and their performance comes into focus.

Avoid the "Dummy Camera" Syndrome

Scarecrows were (and perhaps still are) commonly used to ward off disruptive birds from foraging for food in crop fields. I have seen lifelike statuettes of birds of prey used as a deterrent to keep pesky birds away from places or business (e.g. waterfront restaurants) or where flocks of birds pose a danger, such as at airports. But once a few clever birds determine that there is no risk from the artificial measures, the tactic has failed. Likewise, organizations must not be fooled into thinking that dummy security cameras and empty threats will thwart fraudulent behavior for too long: Once employees or other potential fraudsters realize that there is no or very minimal real danger of getting caught, the organization will have left itself open to fraudulent behavior. Worse, the organization may continue to believe that its fake tactics are working while its employees are stealing from it, leaving management blind to what is really going on.

Do not be fooled into thinking that your organization can be fraud-free: Each time someone builds a better mouse trap, Mother Nature breeds a better mouse. People are clever and many well-established frauds resurface or are repackaged with a technological twist. Fraud detection methodologies must be real, with the results tangible and measurable to ensure that they are effective and not disruptive.

The Nigerian letter scheme is an example of a fraud that has been improved upon through the use of technology. A common fraud, as noted by the US Federal Bureau of Investigations (FBI), this combines impersonation fraud with "a variation of an advance fee scheme in which a letter, mailed from Nigeria, offers the recipient the 'opportunity' to share in a percentage of millions of dollars that the author, a self-proclaimed government official, is trying to transfer illegally out of Nigeria. The recipient is encouraged to send information to the author, such as blank letterhead stationery, bank name and account numbers and other identifying information using a facsimile number provided in the letter." Instead of these letters being sent via postal mail, they are now transmitted en masse via email through the Internet.

Checklist for Fraud Detection

- DO engage employees, vendors, and customers in the fight against fraud through education and collaborative efforts such as sharing data.

- DO enable employees, vendors, and customers to anonymously report suspected fraudulent activities via the use of hotlines.

- DO ensure that software controls are used to restrict employee behavior and bear in mind the fact that software controls are only one means of limiting what an employee can or cannot do.

- DO activate software transaction logging features and use the data as an adjunct analysis of employee behavior.

- DO assemble a cross-functional fraud detection team of employees and/or consultants with different disciplines and complementary specialties to handle suspected fraud investigations. DO NOT entrust a fraud investigation to individuals without the necessary skills and experience.

- DO enforce policies governing employee and vendor behavior and be willing to separate the organization from parties proven to have perpetrated fraud.

12

Technology Methodology

Fraud detection technology should enable an organization to go beyond merely gate-keeping and increase the frequency and thoroughness of audit and reviews processes to determine whether fraudulent behavior is present. This can only be accomplished in situations where a trail of evidence is left behind. I have already mentioned transaction logs as one trail of evidence that organizations can use to trace illicit behavior. Fraud detection and reduction methodology coexists very nicely with operational improvement projects. And, like ERP systems, the technologies used to drive operational efficiencies can help an organization to detect and reduce fraud while improving performance and increasing efficiencies. The value of the return on technology and process improvement investments is increased. Justifying fraud detection and reduction projects as stand-alone investments might be a hard sell in some organizations; after all, if an organization has not suffered from fraud in the past, it might be believed by executive management that it is unlikely that it will suffer from such exposure in the future. As myopic a viewpoint as this might be considered, it is not necessarily unusual. For example, people in Florida get lax as years pass between hurricanes and have short memories when it comes to proper hurricane season preparation. Then, at the slightest hint of a hurricane, the home improvement and grocery stores are packed with people gathering the supplies they should have been stocking up on and keeping a good supply of as hurricane season approaches. Making the case for the risk of fraud exposure while presenting on the merits of an operational improvement project might be positioned as the "icing on the cake" to management that has previously shown a reluctance to fund projects solely related to fraud detection and reduction.

Some fraud discovery still relies on the review of paper documentation for clues (for example, fake documents, forged signatures, washed-out ink) to illicit behavior. However, as supply chain partners become more electronically integrated and as more small to medium-size organizations are able to afford and embrace technology, the less paper documentation is likely to exist. It is too time-consuming and costly to audit stacks of paper-based transactions in the hopes of rooting out fraudulent behavior.

Helpful Technologies

At the core of the helpful technologies is the ERP system or similar central business software application or application suite. If an organization does not have an ERP system, the first step is to get one and to make sure it is the right one. The acquisition of individual, targeted software applications (e.g. inventory control) are helpful and can aid in reducing fraudulent behavior, but only within the limited scope of the application's capabilities. To cast a wide net over the enterprise requires software designed for such a purpose – the ERP system – and extending its reach with adjunct technologies. It should be borne in mind that, aside from providing the necessary business functionality, the ERP system needs to have user-level security features and transaction logging capabilities. Security features should prohibit users from accessing information in the ERP system that is not part of their job role and should further define how a user is able to access permissible information, e.g. view versus edit capabilities. Preferred transaction logging capabilities include capturing the before and after value of a data field when it is changed and applying the user identifier, date, and time to all transactions when they are created. The implementation of an ERP system starts to bring control, consistency, and perhaps even some much-needed restraint to an organization's activities, as highlighted in the discussion of supply chain frauds. Some organizations do not have defined business processes or data standards in place, and the effective implementation of an ERP system brings in both. Where organizations manage everything by exception, an environment is created where fraud can easily exist. Defined business processes, consistent data, and transactions with accounting integrity all help not only to create a more efficient business but to allow situations outside the norm to surface more easily. The reason for these exceptions to the new standards might be an attempt to defraud the organization: Be wary of the employee who puts up a fight to bring visibility to processes and information. The ERP system itself may not be enough to cover the inbound and outbound supply chain activities fully and you may need to look at extending the ERP system to cover more of the basics.

I want to mention a quick point of prudence between using a "self-contained" ERP system versus several complementary and connected enterprise applications (e.g. accounting, sales order processing, distribution / warehouse management). In a traditional ERP system the program code already exists to perform data validations from master tables, update accounting ledgers, and move data from one function to another within the software application. In other words there is no (external) interface necessary between the core functions of the ERP system. In connecting different enterprise applications in a type of "best-of-breed" approach, remember that the connection points represent potential vulnerability gaps where illicit behavior could affect data as it is transferred or made available between applications. This should not sway a decision for an organization to use an ERP system versus several enterprise applications in selecting the best software solution. However, the integrity of these points along the data chain must be addressed and closely monitored to ensure that they are not tampered with. This also holds true for the integration of legacy systems to more modern software applications including web portals: Points in between systems where data is passed are places that could be vulnerable to fraud.

In order to locate two key technologies that will help the ERP system cover a wider swathe of business activities and truly enable our fraud detection capabilities across our enterprise, we need to hitch a ride aboard the TARDIS as the Doctor's companion,[1] stand-in for Peabody's pet boy Sherman for a trip through the Wayback Machine,[2] borrow Mr Wells' Time Machine[3] or use your favorite alternate choice of any other time-traveling device for a journey back in time to rediscover the development and introduction of barcode scanning (in the 1960s) and EDI (in the 1970s).

Barcode Scanning

Barcode development began for use with retail store products and on railroad freight cars. The barcode has become an indispensable technology in the retail and grocery supply chains and is achieving growing use in health care and other verticals. Barcode scanning falls under the category of *automatic identification* (Auto ID) and includes newer technologies such as radio frequency identification (RFID) and Visidot®. Most people easily recognize

1 From the long-running BBC series *Doctor Who.*
2 From the 1959 cartoon by Jay Ward and Ted Key about a dog named Peabody and his pet boy Sherman.
3 From the book *The Time Machine* by H.G. Wells.

the familiar one-dimensional (1D) barcodes that are made up of thick and thin vertical (black) lines separated by narrow and wide (white) spaces representing something of an advanced Morse code: It is not just the dots (thin white stripes) and dashes (thick white stripes) but the timing (a short pause as represented by the thin black stripes or a long pause as represented by the thick black stripes) between the dots and dashes that, taken together, represent a single character. (White reflects light back to the scanner's optics, while black does not.) Different barcode symbologies (languages) use a different combination of spaces and stripes to represent a character. There are also two-dimensional (2D) barcodes used on documents, electronic components, packages, and shipping cartons that look like a haphazard arrangement of tiny geometric shapes (squares, rectangles, and triangles). 2D barcodes are great for packing a lot of information in a limited amount of space. Just like 1D barcodes, there are several symbologies of 2D barcodes as well.

In a traditional supply chain role, Auto ID – especially barcode scanning and RFID – is used to record movement of a physical item: product, carton, pallet, container, and so on. Key data fields captured when recording movement include the following:

- The identifier of what was moved, such as a product identifier or a serial shipping container code.
- The quantity of what was moved.
- The movement against some identifier, such as a purchase order or work order.
- The location that the item was moved from.
- The location that the item was moved to.
- The disposition of what was moved (e.g. first quality versus not first quality).

Electronic Data Interchange

The US version of EDI supported by the American National Standards Institute (ANSI) and the more internationally used United Nations Directories for

Electronic Data Interchange for Administration, Commerce and Transport (UN/EDIFACT) are both forms of what can generally be considered electronic business-to-business (eB2B) commerce communications. Instead of shuffling paper back and forth between a customer and a supplier, trading partners can exchange business documents (e.g. purchase orders, invoices, payment remittances, bills of lading, forecasts, sales information, and routing instructions) electronically through the use of data files formatted to a specific standard. eB2B commerce is being extended beyond traditional supply chains (e.g. retail and automotive) to grocery, electronics, health care, pharmaceuticals, marine, book publishing, and other sectors. In some supply chains there are also electronic business-to-consumer (eB2C) transactions that take place and can include the initial consumer order and order fulfillment (order confirmation, shipment notice, shipment tracking) information.

Shared Characteristic of Auto ID and eB2B

The singular most important shared characteristic – at least in terms of this topic – of Auto ID and eB2B commerce is that they can significantly reduce – if not fully eliminate – paper-based business transactions between two parties. Acting as extensions of the ERP system, Auto ID and eB2B enable product (raw material, component, and finished good) movement and supply chain trading partner (customer and supplier) interactions to be performed electronically and driven from the core data repository (the ERP system or collection of enterprise applications). Driving product movement and trading partner interactions away from being paper-based and toward being conducted electronically enables the organization to cover critical information gaps in supply chain operations where fraudulent behavior could occur. Programmatic cross-checks of data are now able to be accomplished more quickly and thoroughly. There is a greater chance of stopping fraudulent activities at the source or at least before they travel too far down the supply chain and manifest themselves into something far worse through better data analysis across different supply chain links that could represent different organizational departments.

Extending an ERP system or enterprise application by implementing Auto ID or eB2B creates the same data vulnerability points already mentioned as those between different enterprise applications. The injection of bad data or the interception and modification of existing data needs to be managed at all points that connect enterprise applications and technologies that extend the software's reach.

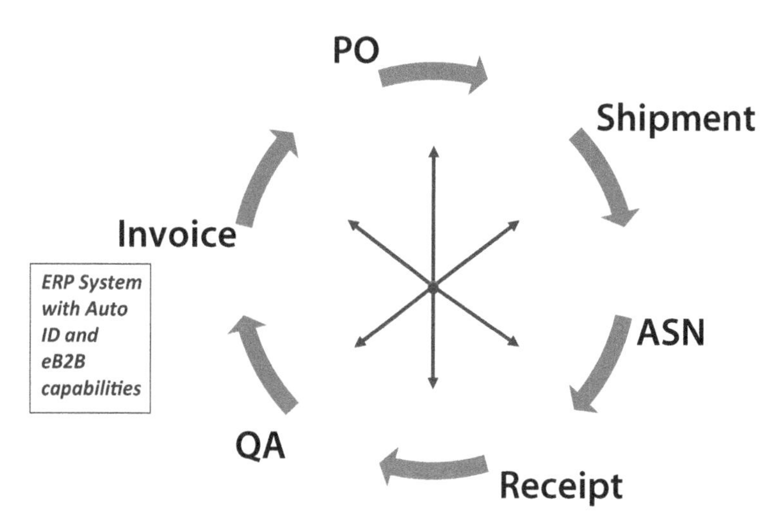

Figure 12.1 Closing Data Gaps

Closing the Data Gaps

With an ERP system in place and integrated with Auto ID (assume barcode scanning for now) and EDI connectivity (assuming this is taking place in the US), a simplified supply chain interaction between a customer and a supplier would resemble something like that shown in Figure 12.1.

In this supply chain scenario:

1. The customer issues a purchase order that is sent electronically (via EDI document number 850) to the supplier.

2. The supplier picks, packs, and ships the product.

3. The supplier sends an Advance Ship Notice (EDI document number 856) to the customer. (The Advance Ship Notice is an electronic bill of lading.) The US retail industry guideline is that the supplier's Advance Ship Notice will be sent not later than one hour after the shipment leaves the supplier's facility or the facility of the supplier's outsourced distribution company, third-party logistics provider, etc. The customer uploads the EDI 856 information to its ERP or warehouse management system (WMS) in preparation to receive the shipment.

4. The customer receives the physical shipment by scanning the barcodes on the pallets and/or shipping cartons or on the items themselves. These serial numbers should match the same data in the Advance Ship Notice.

5. The goods are quality inspected.

6. The supplier sends the invoice (EDI document number 810) for payment.

With the ERP system at the center and traditional paper-based transactions (purchase order, bill of lading, receiving document, quality assurance test results, and invoice) now paperless, the supply chain performance and fraud detection possibilities include the following.

ELECTRONIC PURCHASE ORDER

The catalyst event in a supply chain is the need to obtain something based on a demand, and the associated transaction is the electronic purchase order that authorizes the acquisition of something from a supplier. The purchase order should be generated by an authorized employee based on the organization's needs and not somebody's wants. This means that purchase order amounts should be based on characteristics such as consumption, safety stock, and supplier lead time. User-level limits based on job role in the organization chart should be placed per purchase order and across time periods like calendar or fiscal quarters whereby exceptions require management authorization. This concept can be expanded to the purchase of not just goods or materials but also services, as well as the distribution of funds. The ability to commit funds on behalf of the organization needs to be controlled based on a set of criteria or parameters that places limits based on the business need. Suppliers can be informed that only EDI-type purchase orders are to be accepted and that any attempt to purchase via email, fax, or telephone should be rejected unless absolute proof of authorization can be established. This confines a potential fraudster to the submission of purchase orders through the ERP system and thus subjects the transactions to system controls and limitations.

PURCHASE ORDER AND ADVANCE SHIP NOTICE

The customer hopes the supplier correctly picked, packed, and shipped based on the products and quantities on the purchase order. (The customer *hopes* the supplier correctly picked, packed, and shipped because the customer has no control over the supplier's processes or technology. The customer can trust the supplier but should also verify the supplier's actions.) The Advance Ship Notice electronic transaction representing the paper bill of lading informs the customer of what the supplier states it is shipping. An initial cross-check between the purchase order and the Advance Ship Notice will inform the customer of any potential problems, which may include the following:

- Over-shipments.
- Short-shipments.
- Incorrect shipments (products not on the original purchase order).
- Duplicate shipments (not reflective of a back-order).

Organizations should ascertain whether a shipment will be a problem *before* it reaches the final destination, which may be a warehouse receiving dock or distribution facility or direct to store. Pre-emptively detecting problems when a shipment is in transit gives the organization a chance to thwart an internal fraudster's attempts before a fraud occurs.

ADVANCE SHIP NOTICE AND RECEIPT

Electronically gathering the receiving information allows for cross-checking to the Advance Ship Notice to determine if what the supplier said it shipped matches what it physically shipped and what the transportation carrier actually delivered. This can be useful in detecting shortages and discrepancies between the two, which could be the result of fraud, e.g. theft or cover-up.

RECEIPT AND PURCHASE ORDER

As part of the receiving process, there should be a cross-check to the purchase order for anomalies such as quantity variations from what was originally ordered and unauthorized product substitutions.

RECEIPT AND QUALITY ASSURANCE

A quality assurance check after the receiving process helps to filter out bad products (those that are not first-quality) before they infiltrate the organization. It may not be possible to determine at the point of receipt if inbound goods are usable or not without a more detailed quality assurance inspection.

Goods that fail the quality assurance test should be set aside in a quarantine-type area (meaning that they are not available for use), for supplier pickup or supplier inspection on-site, depending on the nature of the goods and any arrangements with the supplier. Photographs and test results will help the supplier to determine the source of the problem and where the problem may have occurred. For example, perhaps a shipping carton was pierced during normal handling due to the use of compromised cardboard by the box manufacturer, and this led to the contents being damaged or otherwise tainted.

INVOICE AND QUALITY ASSURANCE

The organization should only pay for first-quality goods that were received. Linking the quality inspection results to accounts payable can help prevent payment for unsatisfactory products.

The above scenario points to increasing efficiencies in business operations and measuring supplier performance, and these are goals that can be achieved through the implementation of these technologies. Where the return on investment in technologies and operations improvements becomes greater is by using the same investments to both increase efficiencies and reduce fraudulent behavior. In some circumstances the fact that no operational performance baselines exist can make it more difficult to detect fraudulent behavior and differentiate it from poor operational performance. In justifying investments in technology and business process improvements, organizations should look toward other benefits such as fraud detection and reduction. Combining these goals helps to ensure the solutions selected cover aspects shared by and unique to operational performance and the detection and reduction of fraud. Detecting supply chain fraud may require a comparison of the results between several cross-check points to ascertain fraudulent behavior and indicate that an investigation needs to be started.

Migrating from paper-based to paperless business transactions through ERP systems, Auto ID, and EDI can be collectively labeled as Electronic Records Management (ERM). The US government has an ERM initiative with the National Archives and Records Administration (NARA) as the managing partner. As posted on the NARA website, the vision of the government's ERM initiative is: "To effectively manage and facilitate access to agency information in order to support and accelerate decision making and ensure accountability."

Lynn A. Fish at Canisius College authored a paper in 2008 entitled "RFID and ERP Implementation in the Food Supply Chain: Issues, 2008 State of the Industry and Best Practices." This paper gives an overview of the RFID and ERP integration efforts in the US and European food supply chains.

Dr Fish states that as of 2005, "75% of food supply companies manage inventory through disconnected spreadsheet and paper documents." The need to implement ERM is due to risks and liabilities during a food recall, including loss of company reputation and potential liabilities. The benefits of ERM are process improvements and risk management. According to Dr Fish, the US FDA only requires tracking of the immediate source of and the immediate recipient of food, but lot number tracking is not required. The paper continues that even after the US Congress passed the Bioterrorism Act of 2002, of which the FDA has oversight, to guard against data breaches and the infiltration of toxic substances in the food supply chain, there is no ERM requirement. The Act only "requires the establishment and maintenance of records by persons who manufacture, process, pack, transport, distribute, receive, hold, or import foods in the United States."

Dr Fish cites examples of tainted food problems that prompted recalls in the US and Europe. (Tainted food examples include those where an ingredient not listed was introduced into the food and where there was the presence of harmful bacteria or other toxic substances.) She suggests that a lack of ERM results in delays in the ability to identify and recall products as well as informing consumers of the potential hazards. Without ERM, throughout the food supply chain there is a disconnected "chain of custody" and the ability to track from "food-to-fork" is impeded.

Business Intelligence Reporting

Using Auto ID and eB2B, the organization increases the proverbial breadcrumbs along the electronic business transaction trail by moving from paper to paperless processing and thus greatly enhances the ability to cross-check information to look for fraudulent behavior patterns. Care must be taken to differentiate a sloppy supplier's poor performance from that of a scheming supplier's fraudulent behavior. Remember: Fraud is performed with the intent to deceive. A supplier that just doesn't care about its performance is different from a supplier that is deliberately trying to swindle its customer. And do not forget that suppliers (and customers) can be internal entities within the organization. The supply chain scenario above could easily be modified to reflect the relationship between two internal departments or divisions.

With all this data, organizations need business intelligence reporting tools to analyze anomalies in order to determine whether fraud is occurring. Return to the case of the public agency supervisor who siphoned off a significant number of payments all under the apparent $900 limit established in the agency's software application, an audit program that counted the quantity of payment withdrawals (also known as a "draw") and the total monetary amount drawn per employee by month would have been a good start in detecting this fraud (see Table 12.1).

Table 12.1 Employee Draws per Month

Employee	Month 1	Month 2	Month 3	Month Averages
1	80 draws $40,000 total	60 draws $54,000 total	70 draws $59,500 total	70 draws/month $59,503 avg/mo
2	75 draw $46,875 total	75 draws $61,875 total	85 draws $72,250 total	78 draws/month $60,333 avg/mo
3	80 draws $48,000 total	75 draws $48,750 total	80 draws $56,000 total	78 draws/month $50,916 avg/mo

Based on the starting analysis, it does not appear that anything is amiss: For each employee the average number of monthly withdrawals and the average monetary amount seem reasonable when averaged horizontally across the time periods. It does appear that Employee 3 is providing less money per withdrawal than the other employees: The average monthly withdrawal amounts for each employee (1, 2, and 3) are $850, $774, and $652 respectively. However, it is difficult to come to meaningful conclusions from these numbers. Is Employee 1 suspect because his average withdrawal amount is higher? Is Employee 3 being too mean in providing less money per withdrawal than the other two employees? Employee 1's average withdrawal is the highest and perhaps requires more scrutiny.

Table 12.2 brings more information into the picture with monthly averages. A vertical analysis that compares monthly averages to each employee's number of withdrawals and total monetary amounts indicates that Employee 2's withdrawal amounts in months 2 and 3 were noticeably higher than those

Table 12.2 Employees Draws per Month with Averages

Employee	Month 1	Month 2	Month 3	Month Averages
1	80 draws $40,000 total	60 draws $54,000 total	70 draws $59,500 total	70 draws/month $59,503 avg/mo
2	75 draw $46,875 total	75 draws $61,875 total	85 draws $72,250 total	78 draws/month $60,333 avg/mo
3	80 draws $48,000 total	75 draws $48,750 total	80 draws $56,000 total	78 draws/month $50,916 avg/mo
Employee Averages	78 draws/month $44,958 avg/mo	70 draws/month $54,875 avg/mo	78 draws/month $62,583 avg/mo	

Table 12.3 Employee Draws per Month with Case Details

Employee	Month 1 (20 work days)	Month 2 (20 work days)	Month 3 (22 work days)	Month Averages
1	80 draws $40,000 total 4 draws/day 150 cases $266 average draw/case	60 draws $54,000 total 3 draws/day 145 cases $372 average draw/case	70 draws $59,500 total 3.5 draws/day 155 cases $384 average draw/case	70 draws/month $59,503 avg/mo 3.5 draws/day 150 cases avg $397 average draw/case
2	75 draw $46,875 total 3.75 draws/day 125 cases $375 average draw/case	75 draws $61,875 total 3.75 draws/day 125 cases $495 average draw/case	85 draws $72,250 total 3.86 draws/day 130 cases $556 average draw/case	78 draws/month $60,333 avg/mo 3.77 draws/day 127 cases avg $475 average draw/case
3	80 draws $48,000 total 4 draws/day 170 cases $282 average draw/case	75 draws $48,750 total 3.75 draws/day 165 cases $295 average draw/case	80 draws $56,000 total 3.6 draws/day 165 cases $339 average draw/case	78 draws/month $50,916 avg/mo 3.78 draws/day 167 cases avg $305 average draw/case
Employee Averages	78 draws/month $44,958 avg/mo 148 cases/mo	70 draws/month $54,875 avg/mo 145 cases/mo	78 draws/month $62,583 avg/mo 150 cases/mo	

of the average. We might begin to suspect Employee 2, but still the analysis in inconclusive.

Table 12.3 provides additional information that is necessary in performing the detailed analysis required to determine whether an employee's behavior is suspect by including the number of cases each employee had responsibility for during each month in the analysis.

With the additional information of case load, it is possible to obtain a clearer picture of abnormal behavior. All employees averaged approximately the same number of withdrawals per day. Note that Employee 2 had fewer cases during the three-month time period yet paid out a higher per-case withdrawal amount when the averages are compared vertically among all employees. A possible reason for Employee 2's behavior could be that he simply has a kind heart and wants to provide as much help to families in need as possible; however, this may run afoul of agency guidelines. In this possibility Employee 2 might require additional training and education on maximum per-case benefits amounts. Another possible reason for the skewed numbers could be that Employee 2 is collecting a kickback from the beneficiaries in providing higher than average withdrawal amounts. This should warrant a review of Employee 2's cases to determine whether beneficiaries were falsely allowed into the system (e.g. by providing faked financial documentation) because the investigation should also attempt to ascertain whether the beneficiaries themselves are legitimate or not. Nepotism is not an uncommon discovery when illegal or unethical behavior in the areas of purchasing and fund distribution (e.g. for grants) is investigated. Money is a critical component of any type of supply chain and the ability to trace its inbound and outbound flow is a necessity. "Does the business transaction make sense?" is a question that should be repeatedly asked when analyzing the money chain or the financial part of the supply chain.

An analysis similar to that outlined above could be used to review any purchasing function in an organization. For each "what" that is being purchased, how often and how much are two key data points for the analysis as to whether the business transactions make sense. "Why" something is being purchased can be supported or refuted with data representing demand. Taken all together, this information is useful in determining not just whether smart purchasing decisions are being made but if there is any suspicious activity taking place in the purchasing function.

There is a risk that analytical reporting may be flawed, in which case the relationship with the supplier may suffer damage if it is falsely accused of illicit behavior. The organization has the responsibility for ensuring its reporting is as error-free as possible and is meaningful. This requires data set-ups and transaction validations to make certain that there is a solid foundation upon which to base any accusations of illicit behavior. It is important to ensure that goods are associated with the correct vendor and that each vendor is uniquely and uniformly identified throughout the purchasing, receiving, quality inspection, and invoicing processes. In the receiving process, for example, it is vital to ensure that the same carton is not counted twice and to double-check to make sure that a carton does not go uncounted. If possible, it is best to avoid having the same employees quality check the same suppliers time and time again, whether on the receiving dock or in the quality inspection laboratory. Receiving via unique carton identifier against a purchase order can aid in the validation. It is crucial that the organization has complete integrity and accuracy in its processes, data gathering, and analytical reporting before judging – let alone accusing – a supplier of poor performance or fraudulent behavior. Output results might show comparisons against normal or average behavior patterns so that anomalies are more noticeable. As indispensable as technology is to the detection of fraud, the necessity of human judgment does not go away.

13

Due Diligence in Detection

Detecting supply chain fraud means that we have to look for it even when we are not faced with any evidence of wrongdoing. Regardless of all the data gathering and analytical reporting, procedures, and controls, there is a necessary due diligence that organizations must perform: Sometimes it is making sure that nothing is missing or that something else is not present, or that there is not too much of one thing or not enough of something else. You won't always know what you're looking for, but look you must. Engage employees at all levels and ask their opinions about business processes, technologies, and what they experience on a daily basis.

When I speak to various audiences on the topic of supply chain fraud, I tell them that while an organization can outsource something like a manufacturing process, it cannot outsource the responsibility for what it has outsourced; it bears the responsibility itself when it faces its customers if something goes wrong. It is not a good assumption that just because there is a contract to purchase, this equates to a contract of compliance. The buying party has a responsibility to perform its due diligence to ensure to a reasonable level that the goods it is ultimately responsible for are safe for use (which may include consumption) by its customers. The definition of a reasonable level of due diligence may be defined by regulatory laws or industry standards, or it may be left up to the organization to ensure that the best efforts (and yes, "best efforts" would require further definition) are undertaken to establish the safety or wholeness of its products or services.

Consider the following stories of tainted product problems.

Automobile Tires

In June 2007 a tire importer in New Jersey was forced to recall 450,000 Chinese-made tires for separation problems. It seems that there was either no gum strip between the belts of the tires or that the gum strip was not up to standard. Without sufficient bonding between the belts normally provided by the gum strip, the tires would separate under normal use. This incident resulted in several deaths due to tire failure. This case had similarities to the Bridgestone/Firestone recall of 2000.

Milk, Toothpaste, and Pet Food

In 2007 it was discovered that Sanlu – a Chinese dairy company – was selling milk and dairy products tainted with the chemical melamine. As summarized in an article entitled "Food Fraud in the Global Supply Chain" in the March 2010 issue of *Food Logistics* magazine, the introduction of melamine in Chinese-made milk and dairy products had a $10 billion global impact. Somwhere in the order of 290,000 people were made ill, nearly 52,000 people were hospitalized, and approximately six died. Over 60 countries initiated bans or recalls of affected milk and dairy products. Interestingly, it does not appear that the Chinese government had set standards for acceptable levels of melamine in food or beverage products until around October 2008, well after this crisis had erupted. This case has similarities to the problems that resulted in a US FDA warning in June 2007 regarding toothpaste manufactured in China that was tainted with an ingredient normally found in automobile coolant known as diethylene glycol (also known as diglycol or diglycol stearate). In 2007 melamine was also added to Chinese-manufactured pet food that was shipped to Menu Foods in Canada for (private-label) preparation for distribution into the US. Approximately 60 million cans and pouches of pet food were involved in the recall.

Toys

In 2007 the US Consumer Product Safety Commission (CPSC) issued several recall notices for millions of Mattel and Fisher Price toys (manufactured in China) that were found to contain lead parts or lead paint. Mattel was subsequently fined $2.3 million in a civil penalty aside from any lawsuit settlements (see www.mattelsettlement.com). According to a June 5, 2009 article in *USA Today*, the companies denied willfully violating the 30-year-old federal ban on lead in toys. In a related story reported in early 2010, Chinese toy manufacturers began substituting cadmium for lead. (At the time the decades-old ban on lead in children's toys was in place, but there was nothing stated about cadmium, so the toys could legally be sold in the US.)

So, what do these tainted product cases have in common?

1. The products were all manufactured and/or distributed in 2007.

2. The products were all made in China.

The logical conclusion is that 2007 was not a good year for Chinese-made products. But there is more to this than just the obvious. In all of these cases, why weren't quality assurance checks carried out throughout the entire supply chain and, if they were performed, why didn't they catch any of these problems? In each case the quantity of problematic products is pretty staggering and statistically significant, such that it is difficult to believe that no one spotted any of these problems before the products found their way into the hands of consumers. And yet the truth seems to speak for itself: Either no one looked for these problems or someone found them but the truth of the discovery was initially buried.

A January 1, 2009 China Internet Information Center website article states that Sanlu formed an investigative team after receiving numerous product complaints in March 2008. In May 2008 the team reported finding "that the products were contaminated with a non-food chemical rich in nitrogen." The report of this unknown non-food chemical was sent to Sanlu's chairwoman and general manager at the time, and apparently under her orders further investigation revealed in late July 2008 that the non-food chemical was melamine. The article continues: "Sanlu then sent samples from 16 batches of its products to the quality inspection and quarantine bureau of Hebei province to test for melamine. It, however, did not specify that the samples were from its products. The results, issued on August 1, showed 15 of the samples contained high levels of melamine."

What this case shows us is that quality assurance inspections can reveal the existence of an unknown ingredient and that further tests can discover its identity. One hopes that had Sanlu performed quality assurance tests in the early stages and throughout the production process – as well as on the finished goods before distribution – there is a good chance that none of the tainted product would have been sent to market. I say a "good chance" because it is obvious that fraud did take place: The melamine was added to make up for a sub-standard ingredient or to "water down" the base product. Assuming the executives themselves and the majority of the hierarchy of management underneath were not part of the fraud, it is to be hoped that had test results of a foreign ingredient been known, the product would not have been shipped.

Let us assume for a moment – just for the sake of argument – that the entire Sanlu organization was guilty of fraudulent behavior and knew of the tainted products it was shipping. It was not just milk and finished dairy goods but also dairy-based ingredients that were used by other organizations (e.g. Cadbury, Nestlé SA) in the production of their products. Therefore, how could the recipient organizations not perform quality tests on the food and beverage ingredients they purchase from a third-party supplier? (Not that it really makes a difference whether the supplier was a third party or not, as mentioned before.) The same question can be asked of Mattel: How could this company not quality test to ensure compliance with a (at the time) 30-year-old federal ban on lead in toys? And should it not have been checking for any unknown substance (e.g. cadmium) in the toys it was purchasing from its third-party supplier? As summarized in a September 21, 2007 article in *Time* magazine and on its website, Mattel admitted to a lack of quality inspections at critical supply chain points and apologized to its Chinese manufacturers for placing the blame on them for the lead problems as well as what turned out to be design flaws that allowed powerful magnets to come loose and be swallowed.

The mortgage crisis in the US spurred banks and mortgage-lending financial institutions to charge ahead with the filing of foreclosure lawsuits. These financial institutions outsourced these services to law firms such as those of David J. Stern, a Florida attorney who once employed nearly 1,200 people at his firm DJSP Enterprises, which at its peak was processing nearly 75,000 foreclosure cases per year. The firm's collapse was due in part to "allegations of fabricating documents, slipshod paperwork and questionable fees" as reported by the *South Florida Sun Sentinel* newspaper.

As reported by MSNBC and CBS' *60 Minutes* news show, financial institutions themselves hired unqualified persons to act as "robo-signers" in order to mass-process foreclosure documentation. According to MSNBC: "Financial institutions and their mortgage servicing departments hired hair stylists, Walmart floor workers and people who had worked on assembly lines and installed them in 'foreclosure expert' jobs with no formal training ... Many of those workers testified that they barely knew what a mortgage was. Some couldn't define the word 'affidavit.' Others didn't know what a complaint was, or even what was meant by personal property. Most troubling, several said they knew they were lying when they signed the foreclosure affidavits and that they agreed with the defense lawyers' accusations about document fraud." According to the *60 Minutes* investigation, financial institutions were outsourcing the foreclosure documentation processing to firms that were hiring high school students, and these students were found to be forging signatures of financial institution vice presidents at the rate of more than 4,000 forged signatures per day.

Outsourcing is not limited to manufacturing or assembly: As the mortgage crisis has revealed, outsourcing can include the processing of financial documents. One question the financial institutions should have been asking is whether the throughput of the firms it was outsourcing to made sense. It does not appear that any financial institution benchmarked the amount of time a moderately experienced person would require to read through, audit, and process mortgage loan documents in preparation for filing a foreclosure. This benchmark could have been used to determine whether an outsourced law firm or even the financial institution's own internal departments – based on the amount of personnel – were spending sufficient time before filing a lawsuit that would essentially remove a person from his or her home.

Corruption and Corporate Fraud

The *Merriam-Webster Dictionary* (www.merriam-webster.com) defines corruption as both an "impairment of integrity, virtue, or moral principle" and an "inducement to wrong by improper or unlawful means (as bribery)." In other words, corruption represents an important type of fraudulent behavior: Confidence is breached due to an impairment of judgment in all likelihood resulting from an inducement of some kind. Corporations are considered corrupt when their leaders are engaged in illicit activities and, in many cases, the captain (sometimes with some help) takes the entire crew down with the proverbial ship.

Were these tainted product and service scandals the result of corruption by rogue entities (employees and/or people outside the organization) akin to occupational fraud or were they more indicative of corporate (organizational) fraud? It is worth considering that both occupational and organizational fraud could involve a combination of any or all of the following illicit activities and conspiracy scenarios:

1. Bribes or kickbacks – typical unlawful means of inducement.

2. Employee collusion with a competitor to damage the organization's reputation – such action could help drive business to the competitor while the organization is mired in damage control and rebuilding the perception of its character.

3. Employee collusion with a supplier who cannot provide first-quality product – possibly what happened with some of the tainted product frauds.

4. Employees acting on their own – a rogue employee could have sold some of the ingredient inventory, pocketing the ill-gotten gains, and then tainted the remaining ingredients or otherwise used some substituted product (a replacement or a "watered-down" mixture) to compensate for the missing ingredients in whole or in part. Alternatively, an employee might bypass certain oversight steps in order to save time and boost throughput.

5. Anger or revenge by an employee against the organization carried out in order to harm the organization's reputation – such actions can be set off by a bad performance review, being denied a raise or adequate bonus, or being passed over for a promotion.

6. An attempt – internally or externally – to cause the organization's stock price to be negatively affected by a competitor or perhaps a potential investor.

From my point of view, I see a distinction between those behaviors perpetrated by disenchanted employees and that of enterprise negligence, which is more indicative of corporate fraud such that gross negligence in detection is an executive management failure. In these and other tainted product and service scandals, the necessary quality assurance inspection resources were either non-existent or reduced to being ineffective as a result of cost-cutting moves designed to achieve the following objectives:

a) Reduce operating overheads.

b) Lower the cost of goods sold.

c) Increase the profit margin.

d) Boost stock prices.

e) Increase management bonuses.

f) Retain customer loyalty through low consumer prices.

Return on investment is as important a perspective as risk management, but in the end it is my firm belief that it is far better to be safe than sorry, especially when lives (whether they are those of people or pets) are concerned. In these tainted product and frauds the question I have is simple: Where was quality assurance? If it did not exist or was driven to impotency through excessive cost-cutting so that executive management could either directly or indirectly benefit from a better balance sheet, then there was a failure in priorities and effectively managing risk.

As reported in the November 4, 2011 edition of the *South Florida Sun Sentinel* and attributed to Reuters, the world's largest maker of cereal, Kellogg, admitted that excessive cutting of employees led to problems, including issues with food safety. Kellogg stated that the company would spend $70 million in the second half of 2011 on manufacturing improvements, which included hiring 300 people to add to the company's factories. In June 2011 listeria was found at a Kellogg factory by US regulators, who warned Kellogg about "significant violations" of regulations governing good manufacturing practice. John Bryant, Kellogg's current CEO, was the chief operating officer when much of the cost-cutting was executed. Bryant stated that Kellogg "cut deeper than it should have" and added: "We'd all rather do that without having to spend $70 million, and that $70 million was a surprise to [Wall Street]; but I think it was absolutely the right thing to do."

The fallacy of believing that what happens outside the organization requires little oversight reminds me of the early days of outsourced programming: The failure of organizations to create thorough technical and operational design specifications led to software code delivered that failed to function as necessary or could not be (easily) modified as business requirements changed. Businesses finally learned the hard way (though it was quite obvious to some of us) that outsourcing can work but that it requires dedicated management and clear communications not unlike what is required internally. This is something that has really always been required but is too easy to often overlook and cast aside as a time-consuming waste of resources. (Isn't it better to do things right the first time than to have to do them over again?) For all the quality management that organizations put into software development, it is simply unacceptable to do anything less when it comes to the products that the organization makes.

Due diligence in fraud detection requires top-level management awareness that the supply chain is vulnerable at so very many different points. It is only through dedication to proactive risk management that supply chain security and integrity will be accomplished, with the result being a safer overall supply chain, right down to the consumer's use of the organization's products.

Getting Started on Fraud Detection

When faced with the daunting task of getting a handle on an organization's business processes, I find it best to start with a picture of the way in which the organization currently functions. A good way to begin is by creating a flowchart of the inbound and outbound supply chains. Each flowchart link represents a place where a process occurs, such as entering a sales order, creating a purchase order, generating an invoice, etc. At each link, it is important to identify the person or job role responsible for the action and note which tools are utilized in the performance of task, such as the software application form being used.

The links on the flowcharts may be represented by other flowcharts because they are processes themselves. Do not let the flowcharts become complicated, which can happen if more than one activity or concept is represented, e.g. receiving goods versus picking against a sales order. Let each flowchart represent a single activity. Higher-level flowcharts should "call out" lower-level flowcharts that detail a specific process. As the flowcharts are being created, all process paths must come to a satisfactory conclusion, such as a document (e.g. purchase order) being sent to a supplier, or a data handoff point between two departments or between two people within the same department. It is important not to let a process path end without something decisive happening. Gaps and loose ends are not just where operational inefficiencies can reside but where fraudulent activities can lurk. As the current state of the business operations is defined, the deficiencies will become more obvious. Walking through each process with test data will help to ensure that all scenario variations have been accounted for in the process flow and that nothing has been missed. Continually refining the current state process flows will help to ensure that exceptions to the norm have been satisfactorily addressed (ether eliminated or provided their own process paths) and that the normal operations flows are designed to cover operational and software-related gaps. One of the benefits of this exercise is that what an organization does not know is likely to be revealed. Not knowing what risks exist is a risk in and of itself.

14

Fraud Ramifications

As part of good business management practices such as providing motivation, bonus incentives, educational opportunities, growth strategies, and career advancement pathways, risk management conversations should have taken place where the cost of supply chain frauds was weighed against the value of good detection and reduction programs. Either no one brought the potential vulnerable points to the table or executive management failed to pay them the necessary attention. Ultimately, the board of directors and officers are responsible to the organization's stakeholders for the costs and other consequences of supply chain fraud, and therefore it is their job to ensure that systems of detection and prevention are in place, both for fraud and corruption. The board of directors must ensure that corporate officers are focused on fraud detection and reduction in private enterprises, public companies, and government agencies charged with the responsible use of taxpayer monies.

The ramifications of supply chain fraud have wider implications which can adversely affect other parts of an organization or its processes. A thorough risk analysis should consider effects such as the following and could be factored into any cost analysis to justify implementing a fraud detection program beyond merely the need to ensure the production of a safe product or providing a safe service.

Manufacturing Downtime

Consider the examples where melamine was added as an ingredient to edible products. In all likelihood, affected food processing machines will have had to be cleaned to rid them of the harmful chemical. This downtime may have required the disassembly of machines in order to perform the necessary cleaning and sterilization. Repairs to internal workings and the replacement of components may have been required if the chemical caused damage.

Customer Credits

The manufacturer or distributor may be required to issue credits to customers (e.g. retailers, distributors, manufacturers) based on the sale of a harmful or non-first-quality products, which includes finished goods and ingredients. The credits may include the cost of the product as well as shipping fees if those were paid by the customer.

Product Replacement

The manufacturer or distributor may be contractually liable for product replacement. This could force a manufacturer toward extra shifts and overtime to make up for the extra capacity needed and to still fulfill other customer orders in the pipeline. The manufacturer may be required to express-ship products to meet this sudden production demand.

Consumer Liabilities

Lawsuits may arise from consumers due to injury or death from the use of a tainted or unsafe product. Lawsuits may also be initiated by individual state attorney generals (applicable in the US) or by the federal government.

No Products to Market

It may be difficult from a public relations standpoint for an organization to release new products to market in the wake of existing products that are the causes of problems. Delays in bringing new products to market can result in lost potential sales and higher inventory holding costs for raw materials and the finished good itself. The new product might need to be reworked to (further) distinguish itself from the product that caused the problem.

Advertising Campaign Delays

Any advertising campaigns for current or new products might need to be cancelled due to the fact that there are harmful products in the marketplace. Costs brought about to the creation of advertising campaigns may be lost if the campaign cannot be reused in the future. Secured (pre-paid) advertising space may result in full or partial losses depending on the contract with the media outlet.

Damage Control Advertising

The organization will likely have to run damage control advertising – possibly in excess of normal advertising campaigns – in major media outlets (newspapers, magazines, television, and the Internet).

Vendor Compliance Chargebacks

Beyond issuing customer credits, the manufacturer or distributor may be liable for vendor compliance chargebacks, which are financial penalties assessed by the customer for compliance violations to the vendor (supplier) contract. Vendor compliance chargebacks are very common in the US retail industry.

Product Recall

Disposal of the harmful product is likely to be a cost that is ultimately borne by the manufacturer or distributor rather than the retailer, though the retailer may have to bear at least some initial expenses. There are several cost components of product recall, which include the following:

- Collection – the tainted product must be removed from store shelves or returned by the customers.

- Shipping – the cost of transporting the tainted product to one or more collection points.

- Reworking – if the product is salvageable with proper rework, it may be possible to sell it for a reduced price, although not necessarily back into the same country.

- Destruction – the product may simply have to be destroyed (buried or burned) if it cannot be salvaged. However, this poses environmental problems as the product might need further processing to neutralize harmful chemicals or remove dangerous components, such as those in electronic devices made of toxic metals.

Loss of Brand Trust

Also known as an organization's "goodwill," brands have value and brand values can help bolster stock prices. Conversely, brand value can be negatively affected due to supply chain fraud, which can cause stock prices to fall.

Loss of Market Share to Competitors

Regardless of whether the organization provides raw materials, components, finished goods, or services, end users may flock to competitors for their needs rather than continue to place trust in an organization suffering negative publicity. This may be required as end-user entities have their own reputations, stock prices, or safety to consider.

Regulatory Investigations and Audits

On top of everything else, an organization may find itself being investigated by governmental regulatory agencies. This will certainly require support from key employees throughout the organization and thus will detract from regular job responsibilities and current projects, in all likelihood necessitating overtime and possibly the acquisition of temporary staff. Similarly, being found liable for fraud or corruption in one area of the business may prejudice an organization's chances in any future government tenders, and not just tenders for similar business.

In February 2009 the owner of Peanut Corporation of America (PCA), Stewart Parnell, refused to answer questions from a Congressional committee with regard to his company's responsibilities for the recall of over 1,800 products as a result of its failure to halt shipments of peanut-related products that were shown to have been tainted with salmonella. As reported in the news, Parnell and PCA had been shipping salmonella-laced products since 2006 or 2007. According to one email purportedly sent by Parnell, the former company owner directed that unsafe products should be shipped because the safety tests were "costing us huge $$$$$$." According to its website, the company has ceased all operations and is in bankruptcy. The PCA case is an example of supply chain fraud perpetrated by company management who were fully aware of their actions. Here too, the PCA case is an example of supply chain fraud that caused an entire company to shut down, costing the innocent and honest employees their jobs as well.

> As reported in the August 25, 2010 edition of the *South Florida Sun Sentinel* and attributed to CNN, the US FDA is leaning toward the criminal prosecution of corporate executives of food and drug companies involved in product recalls that endanger public safety. According to Lewis Grossman, a professor of law at American University, the regulatory agency has not been aggressively punishing companies for violations that resulted in product recalls. The FDA's authority comes from the "Park Doctrine," which gives the agency the ability to seek convictions against executives for violations of the Federal Drug and Cosmetic Act. Even if the executives are not aware of specific manufacturing violations that compromise quality, if they have the authority to take preventative or corrective actions but do not do so, they risk being branded as guilty and can be charged as such under the doctrine.
>
> The FDA has had the authority to go after executives in this manner, but "the FDA has generally avoided resorting to court to enforce their actions" according to Grossman, who stated that the FDA brought over 600 criminal prosecutions against companies in 1939 but only around 16 in 1989. In 1991 the FDA formed the Office of Criminal Investigations, but the convictions were mostly for fraud and were not directly related to food and drug product safety. The FDA intends to use the Office of Criminal Investigations to enforce Park Doctrine violations. In Grossman's opinion, the FDA's tougher stance is due to the fact that "working it out with the company and the company doing a voluntary recall is not enough" with regard to punishments for an increasing number of food and drug product recalls. Statistics generated by the *Gold Sheet*, a drug quality trade publication, in an analysis of FDA data shows that there were 1,742 in drug recalls 2009 compared to 428 in 2008.

On March 27, 2007, the then US Attorney John Brownlee drafted a statement posted on the US Department of Justice website regarding the guilty plea of ITT Corporation for the illegal transfer of classified night vision technology to foreign countries. The opening paragraphs read as follows:

> *It is the responsibility of the United States government to provide security to the American people. A strong military, equipped with the most modern and technologically advanced war fighting systems,*

is critical to this effort. One of the most significant advantages of our military is its ability to effectively maneuver and fight at night. Enabled by the most advanced night vision technology in the world, our military can see and defeat the enemy in the most severe conditions of limited visibility. Keeping this important technology safe and secure, and out of the hands of our potential adversaries, is critical not only to our national security but to the safety and success of soldiers and Marines in combat today. The secrets of our night vision technologies must be protected.

Today, we announce that ITT Corporation, the 12th largest supplier of sophisticated defense systems to the United States military, will plead guilty to two felony charges, pay $100 million in penalties and forfeitures, subject itself to independent monitoring and an extensive remedial action program, and acknowledge that it illegally transferred classified and/or sensitive night vision technology to foreign countries – including the People's Republic of China – in order to reduce its costs and enhance its financial bottom line. By illegally outsourcing the production of some of the most sensitive pieces of the night vision system, ITT has put in jeopardy our military's night time tactical advantages and America's national security. Simply put, the criminal actions of this corporation have threatened to turn on the lights on the modern battlefield for our enemies and expose American soldiers to great harm. For this, ITT Corporation must be brought to justice.

ITT Corporation was found guilty of violating the US Arms Export Control Act and obstructing a US State Department investigation by omitting facts from certain required reports. According to the plea agreement, ITT Corporation will:

a) Pay a $20 million penalty to the US State Department.

b) Pay a fine of $2 million.

c) Pay $28 million to be split among the federal and state law enforcement agencies involved.

d) Pay $50 million in restitution to the victims of its crimes – the US soldiers.

e) Pay for independent staff to monitor its compliance with US laws.

However, the $50 million restitution was initially suspended and in lieu of payment, ITT Corporation has been given five years to develop more advanced night vision technology for US soldiers. (Any money remaining unspent of the $50 million will be paid to the US government under the plea agreement.) The bottom line is that the US military needs night vision technology that is superior to that of its enemies and this fraud transferred key classified information about US night vision technology to (adversarial) foreign powers in a way that compromised the advantage of the US in this respect. The reason provided for the technology transfer was simply to save money on manufacturing costs.

This particular supply chain fraud was due to an organization going outside of the restrictions imposed on its supply chain due to the nature of the product – the night vision technology. Engaging a supply chain partner with characteristics that violate provisions of a contract that exists between an organization and a customer is fraud. ITT Corporation breached the confidence between itself and the US government, and, I would also add, between itself and the American people. Many US retailers state in their vendor compliance contracts that the supplier shall not source from entities that engage in child labor practices: This puts a burden on the vendor to know its supplier.

At the time that the PCA fraud was reported, its tainted products may have been responsible for causing illness to approximately 600 people across more than 40 states and may be linked to several deaths. It could be argued that ITT's greed could result or will result in the injury and death of US solders depending on how readily the illegally transferred technology can be replicated by foreign adversarial forces. These cases represent two distinctly different frauds with similar results and yet very different outcomes.

Unlike PCA, ITT Corporation is still in business and is still winning government contracts based upon the information on its website. As such, it would seem that despite their culpability, some organizations fare better than others when it comes to their punishment for fraudulent behavior. The key differences here might be that the US military relies heavily on the capabilities of its defense contractors, while consumer peanut products can be more readily sourced from other suppliers as a commodity product. PCA had employed approximately 90 people across three states, while – according to its website – ITT Corporation employs over 40,000 people worldwide and is diversified in several vertical industries, including defense.

15

Reducing Fraud through Transparency

In the 1949 movie *The Fountainhead,* based on the book of the same name by Ayn Rand, Ellsworth Toohey (portrayed by Robert Douglas) manipulates the characters Howard Roark, Dominique Francon, and Gail Wynard (played respectively by Gary Cooper, Patricia Neal, and Raymond Massey) in the story of an architect (Roark) who refuses to yield any of his design concepts to please the mediocre tastes of the masses. Toohey, the architectural critic columnist of Wynard's newspaper *The Banner,* takes advantage of Wynard's absences as the business magnate enjoys the distractions of his wealth and power. In concert with his manipulative scheming, Toohey seizes considerable power at the newspaper. He achieves such authoritative control himself that after he is fired by Wynard, the newspaper all but completely collapses as Toohey's faithful legion of writers and editors walk out in support. Combined with the newspaper's unpopular support of Roark in an incident where the besieged architect blows up a building under construction when he learns of unauthorized changes to his design, *The Banner* is also hit by a near-total loss of readers and advertisers. Under pressure from the board of directors of *The Banner,* Wynard is forced to rehire Toohey simply to gain back his lost employees and – to satisfy readers and advertisers – recants its support of Roark's actions, of which he is later acquitted in a court of law.

Ellsworth Toohey is a great representation of a fraudster. Manipulating the people around them by playing on their weaknesses, acting in a pompous and overbearing manner, causing distractions which prevent focus on the facts, taking advantage of leadership absences, and using their own positions to increase their powerbase are all actions undertaken by people who may very well be up to no good. (Toohey continued to do his job writing his column, but his overall behavior was certainly more destructive than constructive to the organization.) However, the persona of Ellsworth Toohey is not solely representative of a potential fraud perpetrator: Sometimes it is the quiet ones that organizations have to be on the lookout for. These are the employees who are in charge of certain functions and are so dedicated to their jobs that they are rarely absent to the point of forgoing vacations. Other employees know them as diligent workers but – if asked – might also confess that they don't know the full extent of their quiet compatriot's job functions or how the work is performed. How then can an organization identify either the real or potential fraudsters from the mass of the employees, each of whom has his or her own unique set of characteristics and quirks?

Fraudsters generally want to or need to have full control over certain business functions and – in today's more modern times than those depicted in *The Fountainhead* – software application access to successfully execute their schemes. A fraudster's fiefdom must be complete with obstructions so as to keep prying eyes away. Taking advantage of disorganization, a fraudster can appear to be a hero or heroine in maintaining or regaining control over a certain once-wayward business function. Such an employee might be well appreciated by his or her colleagues in performing a job role that no one else would want to have. Here the kingdom's walls were built with the help of those who the fraudster is looking to keep out through their apprehension and admiration. It would not be uncommon for Toohey-type fraudsters to surround themselves with unqualified subordinate employees, the fear being that a knowledgeable member of staff might uncover what is really going on.

The owner of a business was lamenting to me his sorrow with regard to the fact that his once-prosperous company was at the time only making half the sales it once was. However, he also told me that he recognized that he had completely lost control of his business to his three primary managers (engineering, accounting, and manufacturing/shipping) who had each built their own departmental dynasties and did their best to block my analysis for operational improvement recommendations. This was such a problem that the business owner stated that he could not trust the numbers on the financial reports. (The accounting manager even admitted to me that he often fudged the numbers! This was done because, he explained to me, the numbers would "wash" over time and he didn't believe the business owner was smart enough to fully understand the intricacies involved in accounting.) The business owner stated that while he knew that he had lost control of his business, he was also afraid of firing his three managers, believing that it would be impossible to pick up the pieces after their absence and maintain business continuity.

A basic truth is that fraudsters do not want transparency in what they are doing because it is through concealment that they are able to be successful. Organizations that create visibility in operational processes can help to create a barrier to fraudulent behavior. Business models like Lean and ISO (International Organization for Standardization) necessitate the understanding and documentation of business processes and should thus make fraudsters very nervous. Here is another example – like the technologies discussed earlier – where organizations can reap increased return on investment from implementation when fraud reduction is paired with the drive toward more efficient operations.

Are popular business models like Lean and ISO focused enough on supply chain fraud to be effective on their own? A search I performed in May 2010 and repeated in November 2011 on the ISO website (www.iso.org) for the term "fraud" yielded the word's mention in only two of ISO's more than 18,000 published standards. Lean practitioners are focused on improving operational performance by removing wasteful steps that cause inefficiencies. (The existence of fraud can cause operational inefficiencies too.) So, while Lean and ISO are notable for what they are able to achieve, I do not think that they can stand alone as business models in the fight against fraud; something more is needed.

Good Governance

Finding a single, concise, and reputable definition of "good governance" was a more difficult challenge than I had originally envisaged. Governance is basically defined as the act of governing (not much help there) or exercising authoritative control over something (better). It would seem to follow that the term "good governance" means that authority should be exercised rightly, justly, and well. The United Nations Economic and Social Commission for Asia and the Pacific (www.unescap.org) defines good governance by the characteristics listed below:

- Participation.
- Rule of Law.
- Transparency.
- Responsiveness.

- Consensus Oriented.
- Equity and Inclusiveness
- Effectiveness and Efficiency.
- Accountability.

OK, fine, but defining something by listing its characteristics can still be a little bit lacking. What is "good governance?"

In 2007 I completed a corporate governance program through Tulane University's law school. The guest lecturers included individuals who were at the time the heads of US federal government agencies and CEOs of big US businesses, as well as other seasoned business leaders and practitioners. What I thought was somewhat humorous was that, for all their titles and experiences, the best definition anyone could offer for good governance was simply this: Do the right thing. Whatever the "thing" was, you just had to do it right. As simple as it sounds, I do think that this is probably the best definition for good governance, whether it applies to governing a nation or governing an organization: Do the right thing.

So, we all need to do the right thing. But that water gets a little muddied sometimes, and as humans we are prone to errors in judgment, so some guidelines are there in order to help us stay on the straight and narrow pathway. From my coursework I took away what I believe are several key attributes of good governance:

1. Just because it is legal to do something does *not* make it ethical to do so.

2. Individuals should always exercise *full disclosure* of key information – such as hidden relationships – which might skew or otherwise alter their judgment and compromise their impartiality (especially if they can be positively or negatively affected by the outcome) when faced with a decision.

3. Individuals should always exercise *relative distance* when faced with a decision. If upon full disclosure (and sometimes a person is only left to be judged by one's own self) it is obvious that a person

is too close to a situation to make a fair and impartial decision, that person should step away and let the decision be made by others.

4. Organizations should act with honesty and integrity toward customers, suppliers, employees, and competitors alike.

So we now have a definition of good governance ("Do the right thing") and some attributes that provide a type of moral compass to guide us. But is this enough for us frail humans to rely on? Is combining good governance with Lean or ISO enough to root out fraud from an organization? In a perfect world I believe the answer would be "yes," but we do not live in anything like an ideal world.

Taking what I learned and framing it within my own knowledge and experiences of supply chain management and business operations, I thought of two simple questions one can ask of an organization to help determine how well it is governed:

1. Are your employees empowered to ask for – and receive – the training and education necessary to effectively perform their job functions as outlined on their job descriptions in order to be successful in their job roles?

2. Does your organization treat its suppliers with the same value, honesty, and integrity as it treats its customers?

Too often, employees are placed into positions of failure due to a lack of proper job training by the organization. It is of little wonder sometimes that employees are given poor performance reviews by supervisors or managers who themselves are not knowledgeable on the aspects and necessities of the jobs that their subordinates are trying to perform.

While customers are often treated as royalty, suppliers are sometimes treated with disdain by organizations that fail to dedicate sufficient efforts to their downstream supply chain relationships. And even when that does happen, there is sometimes a class system where large suppliers are treated very differently from small and medium-size suppliers, even to the point where the rules are changed and rewritten on a case-by-case basis. Organizations want customers to pay on time but are not always respective of suppliers who also desire – if not rely on – timely payments from their customers. Remember that often an organization is both the customer and a supplier.

Sarbanes-Oxley

In the US the Sarbanes-Oxley Act (SOX) was passed in July 2002 after shenanigans at a few public companies caused their collapse, resulting in massive job and investment losses. Financial books were cooked, monies were misappropriated and misspent, and figures were sometimes more the stuff of fiction than fact. The perpetrators were people in powerful positions with personalities not unlike the fictional Ellsworth Toohey in their manipulation and corruption of those around them. Public companies have a greater responsibility in the reporting of their finances because the scope of their stakeholders (e.g. employees, customers, suppliers) also includes shareholders and potential investors, each of which can be organizations or individuals. According to the US Securities and Exchange Commission (SEC) website: "The [SOX] Act mandated a number of reforms to enhance corporate responsibility, enhance financial disclosures and combat corporate and accounting fraud, and created the 'Public Company Accounting Oversight Board,' also known as the PCAOB, to oversee the activities of the auditing profession."

Alright – now we are getting somewhere. Corporate governance now has a law behind it that defines how public companies should act. (Private companies should take note too, as the lessons learned are certainly applicable for their protection as well. And private companies looking to go public might need to straighten themselves out a little bit first before taking that big step.) Laws are great for telling us what we should and should not do, but they do not necessarily tell us how to do the things we should be doing to ensure that we are in compliance.

> In June 2011 the US Supreme Court voted 9:0 to overturn a Nevada Supreme Court ruling that stated "voting by public officers on public issues is protected speech under the First Amendment." In its reversal (docket 10-568) the Supreme Court decision upholds ethics laws that prohibit public officials from voting on issues in which they have a conflict of interest. Supreme Court Justice Antonin Scalia stated that conflict-of-interest rules "have been commonplace for over 200 years and they have never been thought to infringe on the free-speech rights of lawmakers." Scalia also stated that while the US Constitution gives a legislator the right to speak freely, it does not give a legislator the right to cast a vote on matters where there is a conflict of interest. He continued that the right of a legislator to vote in a legislature "is not personal to the legislator but belongs to the people."

SOX Frameworks

The SOX Act is not just about the timely and accurate reporting of financial statements – it is also about how organizations get to those end results. The SEC is taking a less and less tolerant stance with public companies that constantly restate financials: It is unfair to the point of being misleading to shareholders and potential investors who base their buy and sell decisions in part on these financial statements. (Lending institutions would also be looking at these financial statements to determine how much money they should loan to the company.) It is important to understand that SOX is also about removing the impediments to inaccurate or delayed financial reporting – it is as much about the process as it is about the results. Therefore, throughout the entire supply chain, there must be integrity in the information, accuracy in the reporting, good conscience in the decision making, and no fraudulent behavior. Business models like Lean and ISO can provide a foundation for removing inefficiencies in operations that can be impediments to accurate and timely financial reporting, but they will not be sufficient without a framework for internal controls and risk management.

In the March 2006 edition of *CFO* magazine, the results of a poll were published where responding public companies indicated which SOX frameworks they used. Respondents could list more than one framework since some frameworks are specific to certain functional areas. The results of the poll are noted below:

- COSO – 82 percent.
- COBIT – 33 percent.
- AS2 – 28 percent.
- SAS 55/78 – 13 percent.
- Other – 2 percent.

It should be noted that some responding public companies listed using more than one framework, thus the total percentages add up to greater than 100 percent.

The two most commonly used SOX frameworks from the *CFO* magazine poll are COSO (Committee of Sponsoring Organizations of the Treadway

Commission) and COBIT (Control Objectives for Information and related Technology by the Information Systems Audit and Control Association). Together, these two frameworks provide excellent guidelines for both public companies and private organizations to follow in order to improve integrity in operations and software applications. With increased integrity comes greater control, more transparency (or at least more visibility), and improved risk assessment, which, all together, can help lead an organization toward reducing fraudulent behavior – either before it starts or soon after the act – and being able to achieve timely and accurate financial reporting.

Committee of Sponsoring Organizations

According to its website (www.coso.org), COSO was formed in 1985 and is dedicated to promoting "more effective, efficient, and ethical business operations on a global basis." What is appealing to me about the COSO internal controls framework is how its five key aspects comprehensively cover business operations.

CONTROL ENVIRONMENT

The control environment is also known as the "tone at the top." It is set by executive management's words and actions and is akin to the moral compass that directs the organization along its path. If every morning the leaders of an organization walked in and declared to everyone "Good morning – let's see how we can screw our customers and suppliers today!" the staff employees would eventually believe that this seemingly unacceptable behavior was not just OK but that it was actually encouraged by management, possibly to the point of necessity in achieving favorable performance reviews, promotions, raises, bonuses, etc.

There are several problems with a poor tone at the top such as this one. The organization may be guilty of illegal – not just unethical – behavior in its dealings with customers or suppliers and might find itself facing civil or criminal charges. The bad behavior encouraged by management may be so enthusiastically mimicked by employees that the staff believe it to be acceptable to screw the organization itself, essentially creating a free-for-all environment where every employee looks out for themselves and what they can individually pilfer. Customers and suppliers on the receiving end of such dishonest behavior might retaliate against the organization with their own brand of deceit out of sheer necessity.

Precedent sets the standard. Executive management has the responsibility to lead by example and set the right tone for the entire organization. Outlining corporate objectives show employees the paths that the organization is looking to venture down, and establishing clear policies and procedures showcase how those objectives will be carried out. It is not possible to predict the challenges that lie ahead, so executive management must spend time preparing the organization to face obstacles to goals with integrity and not treachery.

RISK ASSESSMENT

Risks are all around us and I think there are more and more of them with each passing day. Risks to an organization can originate both internally and externally; they can be related to weather, suppliers, customers, computer systems, sudden trend changes, global events, competitors, employees, fraud, and more. To fail to identify a risk is a risk itself. As unlikely as it is for an organization to be able to identify all risks, it must nonetheless try. How the organization chooses to address each risk depends on an analysis of the risk. Once a risk is identified, it should be assessed based on its attributes, such as:

- the likelihood the risk will occur;
- the frequency of occurrence;
- whether the risk can be prevented or mitigated;
- whether it will be possible to recover from the risk and whether the organization can continue through it;
- the cost of the risk versus the value of addressing it.

I have spent most of my life living in the state of Florida. Hurricane season runs from June 1 to November 30 every year. The frequency of hurricane season is an absolute: It will happen once per year and last for six months. The likelihood that a hurricane will strike the state is dependent upon complex global weather patterns and is very difficult to predict. There is no way we can prevent a hurricane, but, depending upon the hurricane's strength, the damage can be mitigated via window shutters and extra roof truss straps. During Hurricane Wilma (2005), I had no power for 14 days. I always stock up on canned food, water, and personal items in the weeks leading up to hurricane season. As long as my house is standing, I will be fine for at least one week until normalcy sets back in and stores reopen. After landline telephone and cellular service was restored a few days after the hit, I began calling friends around the area and found one up the road from me with power and Internet access around day six. For the remaining time I went to my friend's house to conduct business, taking my backups with me so that I could reference materials as necessary. The experience was very inconvenient overall, but I was able to achieve personal and business continuity during that time just fine. I did not opt to purchase a gasoline-powered generator: Even with power, if there is no Internet access, there is little that can be done these days. I combined my perishable foods in the early days of the power outage with those of some neighbors who also have gas grills and others with generators to power their refrigerators. We had quite the feast for the first few days until we had to dig into our canned rations. I cannot stop hurricane season or the hurricanes from coming but I can prepare myself cost-effectively to weather the storms and get through the after-effects by building collaborative relationships (otherwise known as making friends with some of my neighbors) and stocking up on supplies well in advance.

In the December 2008/January 2009 issue of *Supply & Demand Chain* magazine there was an interesting statistic from a survey by UPS and the Economic Intelligence Unit that found that only 16 percent of senior executives believe their organizations pay sufficient attention to risk mitigation in their supply chains. I am left to ask when those executives considered their supply chains if they were thinking of only external suppliers and customers or if internal operations were included?

CONTROL ACTIVITIES

Call them what you will: Policies and procedures, approvals and authorizations, checks and balances. Without getting caught up in semantics, control activities do just that – they place a limitation on what can be done. An accounting clerk cannot sign checks because the signature stamp is locked away, only the controller has the key, and it is not within the purview of an accounting clerk to sign checks. A buyer cannot purchase more than a certain monetary or quantity limit from a supplier without manager approval, which may be an authorized override in the ERP system. A sales representative cannot commit to discounts greater than a certain percentage or monetary amount without management endorsement.

I created a software application to produce carton barcode labels for a pharmaceuticals client early in my consulting career. The client asked me to review US FDA 21 CFR (Code of Federal Regulations) Part 11, which was in the final discussion phase at that time and deals extensively with electronic signatures. The purpose of the software application was to control the printing of carton labels – only the number of labels necessary based on the batch size were to be printed and excess labels printed had to be recorded and destroyed to ensure they were not used for illicit purposes. If any of the unused labels were permitted to leave the building, they could have found their way onto cartons of counterfeit products to increase the look of legitimacy. Further, no production labels could be printed by the operator for each batch until a sample label had been approved by the shift supervisor and quality assurance inspector. Each person's approval was electronically recorded in the software through tightly controlled and interwoven application security. In fact, so critical was this application that it was isolated by being placed in a secured room, off the organization's network, and all software applications (e.g. Microsoft Office) that could have been utilized by a sophisticated user to directly access the application's files and alter data were removed. The control activities for this system and business procedure included the isolation of the system so as to ensure the integrity of the data and reduce the incidences of additional labels being produced.

INFORMATION AND COMMUNICATION

The communication of information necessary to perform one's job function can be summarized as: "Who needs what and when?" Essentially, this COSO aspect refers to whether the right people are getting the right information at the right time in order to be able to make the right decision. The "right people" can be employees, customers, suppliers, shareholders, senior executives, or the board of directors. In terms of detecting and reducing supply chain fraud, organizations should ensure that the right information is communicated to the right recipient (which could be a person, a pass-through from one software application to another, or within the same software application but just a different functional area), such that possible fraudulent activity is red-flagged and brought to the right person's attention. The organization may use a rule-based software application that sits on top of its business applications (e.g. the ERP system) and via the monitoring of business transactions against rule parameters highlights possible illicit activity. It is worth considering that had such as rule-based engine or business intelligence reporting application been deployed, the department supervisor who siphoned off payments of no more than $900 at a time might very well have been caught a lot sooner in her fraudulent behavior.

Information integrity is a paramount concern. It should be easy to determine if the right information is being delivered on a timely enough basis or not – that is probably fairly obvious. But it is not as easy to know whether the information is accurate or has full integrity (all the required data was gathered: Nothing was missed and no extraneous data was included). Information must come from a trusted source such as an ERP system. In too many organizations, critical information is kept in spreadsheets and those spreadsheets are the "trusted" source. The problem is that it might not be possible to reconcile the information on the spreadsheets with the foundation business application. Inadvertent calculation errors or complicated links can cause misrepresented totals. Filtering mistakes can exclude critical pieces of information. Fraud can be perpetrated by forcibly fudging formulas and simply entering fictitious figures. Without the reports being driven by and generated from a trusted business application whose accuracy and integrity is maintained, the organization is at risk of deceit by trusting unverifiable spreadsheet reports.

In March 2011 I had pleasure of presenting at an anti-fraud conference sponsored by the North Florida chapter of the Institute of Internal Auditors. One of the speakers prior to me was William Owens. Mr Owens was the former Executive Vice President of Finance and Chief Financial Officer of HealthSouth Corporation and spent five years in federal prison for his role in HealthSouth's alleged $2.5 billion scandal that artificially boosted profits. He carefully described how he and his team perpetrated this fraud, and he stressed one key point repeatedly: The auditors – external and internal – were kept away from the company's financial system and especially the general ledger. When the auditors required data, Mr Owen's team would download information and programmatically alter it before delivering it in spreadsheet form, or would manually create spreadsheets with knowingly false information by entering fictitious numbers and modifying calculations to generate the desired results. Mr Owens deflected his internal auditors' attention by sending them on meaningless assignments in the field. The external auditors – when they raised questions and demanded data access – backed off after Mr Owens would complain that their fees were too high and that he might have to put their contracts out to bid. The potential loss of HealthSouth as a customer seemed to outweigh the professionalism of the external auditors by the fact that despite the lack of data access, the external auditors were content to continue to do their job and retain HealthSouth's business versus walking away from the fees.

MONITORING

It does little – if any – good to put into place the other aspects of the COSO framework if they are not continually monitored going forward. The monitoring activity reviews for characteristics such as accuracy, adherence, relevancy, and effectiveness. Any initial implementation of a software application or a business process should be reviewed for these characteristics as some fine-tuning is usually necessary. Additionally, business changes and the organization needs to react to those changes. The monitoring activity ensures that, as the organization moves to realign or adjust to these changes, there are corresponding changes made to its internal controls. Modifications to software applications, the acquisition of new software applications, new policies and procedures, mergers/acquisitions/dispositions of business units, and new sales channels are all changes that can necessitate a review and an adjustment of internal controls.

As business changes, it is critically important to ensure that new information makes its way into reports and is accurately reflected in those reports. There must be adherence to policies and procedures, with breaches investigated and addressed. If employees are not following or skirting around various approval and authorization policies, the organization needs to know this and discover why it is happening. Perhaps the current controls are inhibiting business from being effectively performed and frustrated employees are sidestepping the rules due to the pressure to perform their jobs, let alone achieve results. It is possible that old control activities are no longer relevant for new business situations and need to be updated. It certainly does no good to put in place controls if they are found – for whatever reason – simply not to be effective; monitoring helps to ensure that the control is doing what it is supposed to be doing. Organizations need to ensure that foxes are not guarding the hen houses, and monitoring will help to determine that. Monitoring controls should be able to determine whether control gaps exist where fraud can be perpetrated.

Benefits beyond Compliance

SOX-compliance framework documentation is required and can be considered a "living" document because it reflects how the organization actually operates and must be continually updated as the organization changes – this is the benefit and the complexity. The documentation should be referenced and referred to as needed and can be used as the basis for an employee training program. This is not documentation just for the sake of creating documentation that will be placed upon a shelf and collect dust. You could title this document "How My Organization Works" as it is effectively the owner's manual.

Organizations should adhere to the guidelines of good governance and it would be great if ethical behavior was embodied by all its employees. However, the difference between reality and that utopian world can be quite significant, and organizations need to understand and protect themselves from this distinction as part of a good risk management program, not just because they are required to, and not just because it is beneficial to do so, but because the documentation of policies and procedures can also be a smart thing to do, especially as it can be advantageous if the organization is faced with charges of culpability in fraudulent behavior.

The US Sentencing Reform Act (SRA) 1984, under the auspices of the US Sentencing Commission (USSC), outlines the parameters for the assessment of penalties against persons (effective in 1987) and organizations (effective in 1991) guilty of criminal behavior. It seeks to remove the disparity in punishment for crimes and to create a system of mandatory guidelines for the sentencing of persons found guilty of a wide variety of criminal behaviors. One of the goals is basically to make the punishment fit the crime that was committed.

For individuals, sentences are determined by first assigning one of 43 possible base offenses to the crime. Each base offense is adjusted on of a sliding scale based on certain characteristics such as a monetary amount. The base offense is then factored in with one of six criminal history categories (e.g. first-time offender versus repeat offender) to calculate the final sentence level. Mitigating circumstances might cause the court to adjust the final sentence level up or down, but there must be evidence to support such a departure from the calculated sentence level.

For organizations, the first consideration is that of a base fine being the greatest of:

a) the base fine from the offense level fine table of which base fines can range from $5,000 to $72,500;

b) the gain to the organization that resulted from the commission of the offense; or

c) the loss caused by the organization.

For example, an organization commits an offense of which the base fine is $10,000. The organization profited to the sum of $25,000 but caused losses/ damages of $50,000 in the perpetration of the offense. The organization's base fine would be $50,000.

After the base monetary fine is set, the court can factor in a series of multipliers based on other conditions such as the organization's:

- history of criminal activity;
- violation of any judicial orders or injunctions during the offense, such as orders not to destroy documents;
- cooperation with law enforcement versus attempts to obstruct justice;
- acceptance of responsibility if/when proven guilty;
- admission prior to an investigation or other type of self-reporting;
- documentation of procedures to prevent and detect illegal activities.

Organizations cannot be held completely responsible for rogue employees who are determined to perpetrate fraudulent behavior. But this does not mean that organizations are absolved from their responsibility to put into place effective programs to thwart or detect illicit behavior. Documenting business processes and steps, policies and procedures, authorizations and approvals, and showing that employees were provided the opportunity to read this documentation can help the organization reduce the assessed penalty. The organization should anticipate an audit proving that the documentation does in fact match the organization's operating procedures, so I would suggest basing the documentation on fact rather than fiction.

PART III
Reducing Supply Chain Fraud

16

Beyond the Hack

Among the points to take away from this book so far is that perpetrating fraud does not require hacking a network; it can be easily achieved while inside the organization's protected infrastructure. Computer network penetration frauds are frequently mentioned in the media because they are typically related to the theft of personal financial information such as credit card data or – in the US – social security numbers. None of the supply chain frauds mentioned herein required hacking the organization's network in order to perpetrate them – they were all performed by employees who had access to the organization's business software applications.

From the March 2006 *CFO* magazine poll, the second most commonly used SOX compliance framework at the time was COBIT. Focus areas of the COBIT framework include how the IT department supports the organization's business objectives, user identity and account management, technology asset management, and security and control.

> The May 2010 issue of *CSO* magazine, a publication for chief security officers and others engaged in security and risk management, contained an interesting article entitled "Numbers Game," which profiled several information technology frameworks to help guide organizations through information security risk assessments. The article provided summaries of the frameworks, interviews with actual framework users, and some pros and cons of each framework. The frameworks noted in the article were: OCTAVE, FAIR, NIST RMS, and TARA. These frameworks might work very well as an adjunct to the COSO and COBIT SOX compliance frameworks when addressing specific aspects related to information security risk. Essentially there are a variety of frameworks focused on various levels of enterprise risk management and conceived from different sources (government, industry, trade associations). Organizations have to determine which frameworks are appropriate to guide them based on the issues or needs to be addressed with an understanding that some framework compliance must be aligned with laws such as the Gramm-Leach-Bliley Act 1999, SOX, etc.

Network versus Application Security

For the typical person who goes into an office each day, one of the first morning activities will be to sign on to the organization's computer network. It is the network security that controls a user's ability to access data files, folders, software applications, and devices (e.g. printers) across the enterprise. Hardware appliances and software firewalls are used to keep outside entities from gaining access to network infrastructure as a front line of defense.

Once the employee is inside the protected infrastructure (i.e. has successfully signed on to the network), individual application security will determine whether the employee can access a software program and, if so, the type of access granted. One of the most common – and necessary – software programs an employee will need to access is the ERP system or other similar primary business application. Assuming that an employee is granted access to – for example – the ERP system, application security should also determine what functions the employee is entitled to use and at how the employee can interface with the application at the granted level of access. The list below highlights the various control access capabilities that might be granted to a user depending on the sophistication of the ERP system's security features:

- Add.
- Change/Modify.
- Delete.
- Inquire.
- Print.
- Export.

Common Setup or Common Mistake?

Should a supervisor or manager have the exact same software access capabilities as the employees that he or she manages? We expect the individual at the higher hierarchical level to be able to perform more than their subordinates, but aside from this, should the manager have the exact same rights and roles as his or her staff? Consider the software application I created to print pharmaceutical carton barcode labels. Neither the shift supervisor nor the quality assurance inspector could print labels even after all electronic signatures were recorded; only the user (data entry operator) was able to perform that function. This clear separation of responsibilities differentiated the job roles of the three employees involved in the barcode label process. (The Association of Certified Fraud Examiners recognizes the separation of responsibilities as a key deterrent to fraud.) One of the data transaction logs I incorporated recorded the user identity, supervisor identity, quality assurance inspector identity, date, time, lot control identifier, and number of labels printed for each and every batch print. This provided an evidence trail of all label-printing activity. While the data entry operator was the only user permitted to print labels, the shift supervisor had the responsibility of communicating the lot/batch size and thus the number of labels required. The quality assurance supervisor ensured the data entered by the operator matched the production run information (e.g. the product being manufactured).

Giving a manager the exact same capabilities as his or her staff might be required if the security features of the ERP system (or other software application) are limited in their scope. I submit that it should be rare that a chief financial officer or a chief operating officer is required to perform the data entry necessary to establish a new supplier, a responsibility more suited for the controller, purchasing manager, or a lead staff person under his or her direction. I am not suggesting any compromise to the authorization process, just a distinction between who is performing which job function in the grand scheme. And for organizations that are in the market for a new ERP system or other business software applications, a detailed review of how application security is handled is definitely something that needs to go on the checklist. It is important to remember that the more people a fraudster has to involve in order to perpetrate his or her illicit scheme, the greater the risk of the scheme failing or being discovered. Separating responsibilities, e.g. data entry versus authorization or approval, is a key strategy in the fight against fraud.

Identity and Account Management

Identity management means that all users should have a unique identifier and password that allows them access to the network or the individual software applications. No users should be sharing their identifier and password. The primary reason is obvious, as it would be impossible to tell which user among those sharing an identifier performed certain tasks. Even users who work on different shifts (not concurrently together) should not share identifiers and passwords – each user's identifier and password should be unique. The organization should have a clear and established policy against the sharing or distribution of network or software access identifiers and passwords. It should state clearly in the employee manual – and repeat the message often – that employees should not share their user network and software identifiers and passwords with any other employee, any contractor to the organization, or any entity outside the organization, and that this information should be kept secure at all times.

Account management refers to the rights and roles of a user who is authorized to access a software application. This is not just about how the user accesses data (the rights) as previously mentioned (e.g. the ability to add, change, delete, or inquire), but the alignment of the granted rights to the job role. The question again surfaces: Should a manager have the same data-entry capability of his or her staff? (The answer is no.) Account management is critical because it is affected by changes to job roles.

If a person is promoted from a staff position to that of a manager, the job role changes and – as per the example I have been using – so should software rights. This is not always the case, as too often the employee is simply given greater rights while all the old ones stay intact, or the old rights are retained so that the employee can help during a transition period. Fine – sometimes there has to be a transition period, but it should be clearly defined with follow up at some point to remove the excess capabilities not warranted for the new job role.

Cross-department moves can cause a similar situation that can leave the door open to fraudulent behavior. An employee who has enough access to both accounting and purchase functions could establish a fake supplier and allow fraudulent invoices to be paid. An employee with access to sales order processing and accounting could establish a fake customer identity to whom goods would be shipped and then could falsify invoice payments. The likely excuse for this state of affairs arising is that the employee will need to help out his or her old department during the transition and so must retain certain access rights, but I think this is just sloppy policy. The employee's functional access to his or her former department should be removed and any assistance the employee provides should be in conjunction with the person who has assumed the employee's responsibilities. Thus, all tasks within the software application are performed under the proper person's account identity. This reinforces the separation of responsibilities as the former employee no longer has direct access to data that is not part of his or her new job role.

Control Objective

Pity the poor IT technician whose job it is to set up new employees with network and application security access or make changes as current employees take on the different responsibilities that come with job role changes, lateral moves, and promotions. I have witnessed how much of a challenge it is for the assigned technician to get it right first time. Why? Typically it is because the technician has no guidance on what the employee should truly be able to access and what kind of access it should be. I have seen technicians being called back to a user's desk several times in a single hour to try and satisfy the user's demands or those of the employee's manager. And at some point the frustrated technician simply takes the easy way out and provides more access than should be necessary. The technician is between a rock and a hard place: How should he or she satisfy the business requirements and fulfill his or her job ticket quickly so that he or she can move on to the next technical need?

It is not the technician's job to determine what access a user should or should not have. Input from the business side (such as the manager in charge) is necessary and can provide much direction, but I still believe a critical gap exists here. The true authority for identity and account management should be a collaborative effort between the IT department, the business unit, and the HR department. It is after all HR that retains the job descriptions for all the roles in the organization and those job descriptions are in all likelihood based on input from business unit (department) managers. HR's collaboration with the business unit – supported by IT's expertise in the organization's software applications – should align the job roles with the specific software applications the employee will require and how the employee will be permitted to interface with them. Thus, IT is removed from a decision-making role which is actually outside of its domain. By engaging HR with the business unit, there is assurance that the employee is granted only the necessary tools to perform his or her job as it is described – no more, no less. Any additional software application permissions or changes to how an employee accesses software should be coordinated with HR to ensure that it is documented in the employee's file and that there is continual alignment with the organization's expectations of the employee's job performance with the ability of the employee to perform the necessary tasks to fulfill those expectations.

In 2006 an employee of the labor agency for Broward County, Florida (whose major city is Fort Lauderdale) was found to have absconded with over $2 million, a shocking figure considering that the agency's annual budget is approximately $24 million. (The labor agency is funded primarily by local and state agencies via taxpayers.) At the time of her capture, no one was quite sure how long the employee's fraud had been going on for, though it may have been as long as her 12 years of employment there. Looking into our fraudster's background, we found that she:

- had criminal convictions but lied about her background on the employment form and this was not verified at any time prior to or during her employment;
- did not have more than a US high school education;
- did not make more than around $32,000 per year in wages;
- was living – with her husband, a property manager – in an $842,000 house and, according to public records, owned four additional properties (apartments and houses) in four different Florida cities from where her and her husband's primary home was located.

Had there been a verification of her background for criminal activity, it would have been discovered that under her maiden name she was found guilty of three felonies in the ten years prior to her employment. In the last of the three cases she pleaded guilty to theft of $100,000 or more and received two years of probation as a penalty. However, at the time that the fraudster was hired, the agency's employment policy did not include background investigations.

Our fraudster was no computer mastermind – this fraud required no hacking of the labor agency's computer network. It was perpetrated by a trusted employee from the inside. Our fraudster was not a manager or supervisor; she was an accounting clerk in the payables area with enough software application access to write herself several checks each month for amounts that ranged from $12,000 to $20,000 over the course of a six-year period. It is believed that she wrote a total of approximately 120 checks to herself during the time that she perpetrated her fraud.

Receiving exceptional performance reviews, the fraudster requested the ability to work at night on a flexible schedule that would allow her to care for her sick mother. Left to her own devices, unsupervised at night, with enough software application permissions, and probable access to the paper checks and check signing equipment, this employee was literally handed the keys to the kingdom to perpetrate her fraud. To make matters worse, she was not caught by agency's accountants or third-party auditors; it was a sharp bank teller who brought suspicions to a bank manager who then alerted the labor agency. The fraudster's mistake? Banking at the same financial institution as the labor agency.

The perpetration of fraud against different aspects of an organization's supply chain – whether it is raw materials, finished goods, services, or money – does not require the intelligence of a criminal mastermind, only a knowledgeable person who understands lapses in system controls and gaps in procedures, and is either able to gain trust or obscure visibility to his or her work.

17

Governance and Fraud

The ability to reduce fraudulent behavior requires an understanding of its source and nature. Where is the incentive to commit fraud coming from and what is the issue that perpetrating fraudulent behavior is supposed to resolve? Some of the common reasons why fraud is committed include the following points.

The Haves versus the Have Nots

Aside from merely the satisfaction of pure greed, fraudulent behavior can be the result of the clash between the "haves" and the "have nots." Essentially the "have nots" are envious of what the "haves" have. Due to the inability of the "have nots" to acquire what the "haves" have through legitimate means, they are left with employing illicit activities. This might be due to the desire to keep up with one's friends or neighbors, especially if a person has already moved into a neighborhood that is beyond his or her means. It can also be a necessity if operating a different or larger fraud scheme: One must sometimes put on the right airs to build confidence and attract the right victims (e.g. the innocent wealthy) to scam.

As reported by Reuters in June 2011, as the gap between the rich and the poor grows wider, the less fortunate are less likely to view the wealthy as honest or fair. This conclusion is based on a study by the University of Virginia psychologist Shigehiro Oishi, whose team of researchers analyzed the results of more than 48,000 respondents to the General Social Survey, a survey of Americans between 1972 and 2008. The analysis determined that the bigger the disparity in the income gap, the less likely a person will view a wealthier person as honest and fair, which can lead a person toward a degraded sense of well-being. "We originally thought economic factors might explain this all, but they don't," Oishi stated. "When people see that some people are really in advantageous and favorable conditions while many of us are not, lack of trust toward the system and toward others naturally emerges." The study analysis revealed that the wealthiest 20 percent reported no changes in their feelings of fairness, trust, or general well-being with regard to their income disparity.

Rage Against the Machine

Some people believe that they have been treated so unfairly by a government agency, a business, or an employer that they lash out and take revenge against the entity. Anti-government fanatics who believe the government collects too much in taxes might cheat on their revenue forms and short-change their payments. On occasion, people who believe they have been ripped off by a business are caught later trying to defraud the business. (Two wrongs may not make a right, but for some individuals it is more about the satisfaction of retaliating for being on the receiving end of poor service or a dysfunctional product.) It is easier today for consumers to hit back against a business – and to do so with far fewer consequences than fraud and vandalism – because the Internet provides an endless array of places where people can post their displeasure with a business' products or service.

Employees who receive poor performance reviews and are denied raises – even when the reason is beyond the organization's control, such as a downturn in the economy – might seek to take revenge on their employer by pilfering office supplies or petty cash. It is possible that the employee is truly not performing well and is deserving of the poor review and lack of (adequate) raise in salary, but that the employee does not see it that way.

External Pressures of Society

Family health problems, the cost of higher education, and addictions (e.g. drugs, alcohol, gambling) are some of the external pressures that society can place upon a person. People with an insufficient support structure – a religious institution, close friends, family – might succumb to the pressure and take up fraudulent behavior as a possible way out of their situation. HR departments at larger organizations will sometimes try and help employees by directing them to local assistance programs, but this requires the employee to admit a personal problem to the employer, and this admission can be a very difficult thing to do. The employee does not want to risk his or her job, which would only add immense weight to the burden already being shouldered. The HR department might not be viewed by the employee as being a trusted or trustworthy entity. The employee may simply decide that with just a little bit of fraud now, it will get him or her over the hump, rationalizing that he or she will somehow pay the money back or that the organization is doing so well that it would not miss a little money here and there. A sense of "haves versus have nots" might kick in if the employee's pressure causes resentment when he or she witnesses upper management driving expensive automobiles or hears of executives taking vacations or business trips to exotic places. Certainly, such negative feelings can help push employees to commit fraudulent behavior to solve their own personal crisis.

Internal Pressures of the Organization

Unfair treatment at the workplace can cause the "rage against the machine" syndrome. The employee may be a very hard worker and have a desire to do better but has been placed in a position to fail. Typically internal pressures result from a cascade effect that begins when problems are sent downward because different (higher) levels of management cannot or will not deal with them. The employee whose shoulders the burden finally falls upon may be not be trained properly or have the necessary skill set to complete the task to the satisfaction of the employee's manager. The employee may not be empowered to effect the necessary changes to get the problem resolved. There might be a misunderstanding between the manager and the employee as to what the problem really is. This can lead to a gap between the employee's perception of his or her job performance versus that of the employee's manager. This performance review blindness can result in the employee being negatively scored for a failure to reach certain benchmarks that in truth were unattainable for the employee to achieve, given all the constraints.

Examples of various supply chain pressures that employees regularly face are listed in Table 17.1.

Table 17.1 Supply Chain Pressures

Increase	Decrease
New Orders	Manufacturing Time
Recurring Orders	Picking Time
Picks Per Hour	QA Time
Mfg Throughput	Receiving Time
Receipts Per Hour	Inventory Count Time
QA Throughput	Order Entry Time

I was attending a very interesting presentation on best practices in business operations and walked away with some insight that will always stick with me. During the presentation, the speaker paused and asked the audience: "Who is the last person to touch a consumer's goods?" Many people – including me – shouted out answers such as "the waiter or waitress," "the cashier," "the inventory picker or packer" – which the speaker said were all very good, albeit specific answers – but it wasn't quite what he was looking for. The audience sat in silence – all of our brains engaged – when the speaker stated the answer he was really looking for: "The lowest paid employee."

The speaker continued that it is very typically the lowest paid employee who is entrusted with handling a consumer's goods last. Organizations have the highest of expectations that their customers' goods will be handled with care and accuracy, yet fail to recognize the efforts of – and adequately compensate – the employees who are charged with the final touch-point. And often – as in the case of waiters/waitresses and cashiers – these employees have in-person, face-to-face contact with the customer. These final touch-point employees can be the deciding factor as to whether a customer turns into a repeat customer, and are often on the front-line for solving problems and cooling heated emotions. Is it any wonder that employees charged with such an important responsibility are often frustrated at receiving seemingly demeaning wages and being treated with disdain yet are pressured to perform to perfection in order to ensure total customer satisfaction?

The boiling point is when the pressure to perform against odds and adversity becomes too great, there are no other options on the table, and the otherwise honest employee is faced with a critical decision: Job or no job. Paraphrasing the human response to an adverse situation known as "fight or flight," I have termed this "fraud or flight," and when flight (leaving one's job) is not an option, an employee can literally be forced to commit fraud due to the pressures created by the organization itself.

The question then is who is responsible for the fraudulent actions? While the employee certainly committed an illicit act, I would argue that it was the organization that perpetrated the fraud in which the employee was a victim. Remember that fraud is a defined as a breach in confidence and in this case the organization breached the confidence of the employee with regard to fair conduct. This is not the same as an employee willfully being enticed to become an active participant in a fraud scheme, as was more the case with Mr Owens of HealthSouth Corporation and Mr Whitacre of ADM.

Employees may be unknowingly part of an organization's fraud scheme. Mr Owens of HealthSouth Corporation stated in his presentation that approximately 20 individuals in the organization were part of the team perpetrating the fraud. Employees who were not part of the team and thus were not involved in the construction and ongoing existence of the fraud were simply playing their parts in operating the business, though they were not aware that by doing so they were helping to continue to keep up the façade. HealthSouth still exists today, though as a different organization, and it is possible that some people employed during Mr Owens' time who were not part of his fraud team still work there. Other organizations such as Enron and WorldCom no longer exist, with many former employees being the innocent victims of the organization's corrupt leadership.

Returning to the case of Peanut Corporation of America, we can ask the question as to whether any of PCA's employees besides the owner knew the company's products were tainted with salmonella and, if so, could those employees have done something about it? Mr Parnell, the owner of PCA, strikes me as an Ellsworth Toohey type of individual who may have leveraged his position as the company owner over his employees. If the company was in a small town that provided a significant percentage of the employment there, an employee who sought to confront Mr Parnell may have been ostracized by the community and lost his or her job at PCA, making it difficult to secure other employment in the town.

A friend and collaborator asked me to write an article on supply chain fraud that he wanted to send to his newsletter list. He wanted to see what kind – if any – of reaction he would get to such an article. I agreed and wrote a simple short article on how an otherwise honest employee could be pressured by an organization to commit fraud exactly as I have described above. Oh boy – did we get a reaction! My friend forwarded me the response from a mid-level manager at a $6 billion corporation. Let's call the respondent Doug.

Doug fired back with both barrels approximately two days after receiving the newsletter. He felt the topic of (internally perpetrated) supply chain fraud to be disgusting and stated that any suggestion that an organization could force an employee to commit fraud was completely outside the realms of possibility. He felt that what I was insinuating was simply repulsive and he failed to believe that any employee could be forced to capitulate to organizational pressures and commit fraud.

A few days later, my colleague and I received another correspondence from Doug, but this time it was an apology for his initial response. It seems that his emotions were in charge when he first read my article, but then upon reflection he knew that I was right. He confessed and openly admitted that if placed in a "fraud or flight" situation – that is, if he was forced to choose between providing food and shelter for his family versus being forced to commit an act of fraud under pressure from his employer – he would commit the illicit act in order to protect his family's well-being.

Individuals will draw their own conclusions, but as I read Doug's second correspondence, I realized that he was a person devoted to his family – a good husband and a good father who would do what was necessary to be a good provider and protector. I am very certain that he would initially resist committing an act of fraud if it were forced upon him, but would capitulate when the pressure became too great and it put his job in jeopardy. What would you do in a situation like this?

The Fraud Triangle

In the 1950s a sociologist and criminologist by the name of Donald Cressey developed his theory of the Fraud Triangle, which states that there are three components that must all come together for the commission of fraud to occur: pressure, opportunity, and rationalization. These three aspects have already been presented in the various examples of fraud noted already, but the worldwide recession that occurred in 2009 provides a different perspective on how Cressey's Fraud Triangle (see Figure 17.1) can be applied.

During tough economic times, employees are under a great deal of pressure. They note the layoffs happening around them and wonder if they will be next; they have to continue to fulfill their job responsibilities to the satisfaction of their managers while also being assigned additional workloads to compensate

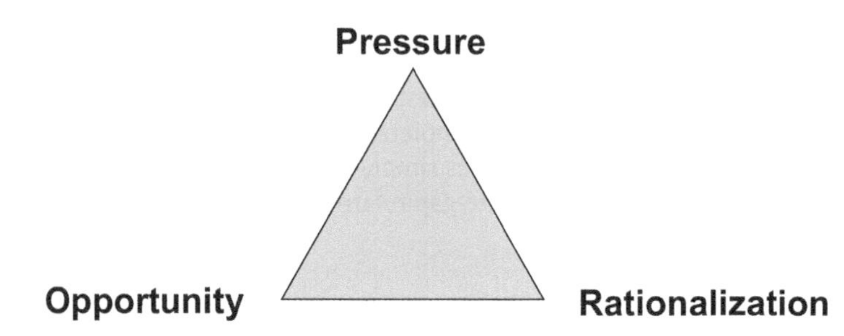

Figure 17.1 The Fraud Triangle

for the workforce reductions. It is very likely that due to slashed budgets, employees will not have training opportunities available to them that would not only advance their skills but would allow them to learn what is necessary to handle their new task assignments. Salary increases are reduced if available at all and bonuses are nil. Combine this pressure with that of everyday life and a person can be left with a severe case of anxiety about his or her current employment situation and what the future might hold.

Job layoffs will result in the consolidation of roles into the hands of the few remaining employees which means these remaining employees must be granted additional business software application access to perform their new assignments. Oversight of internal controls will likely be compromised if job layoffs affect all levels and all departments. In fact, the very internal controls themselves may be compromised, weakened, or diverted around in order to get the necessary work performed with fewer employees. The opportunity to commit fraud has now been greatly increased.

What is left to thwart the commission of fraud is perhaps the frailest of the three components: It is our self and our individual values, morals, ethics, and principles. Shaped by our backgrounds and beliefs, our social circles, and our senses of right and wrong, how we each rationalize our actions may now be the only thing that stands in the way of fraud being perpetrated or prevented. With the door of opportunity open and the crushing weight of responsibility on one's shoulders, it may become easier to rationalize perpetrating fraud in order to relieve the pressure.

Public companies may find themselves caught between a rock and a hard place between meeting the short-term (e.g. quarterly) expectations of financial analysts, boosting shareholder value, and working toward long-term strategies based on investments in research and development, organization goals, market shifts, consumer demand, etc. Notwithstanding the legitimate reasons for lowering overhead costs – including reducing the workforce – organizations must resist compromises to staffing that reduce the effectiveness of internal controls, sacrifice quality assurance programs, or open up the opportunities for fraud without at least putting controls in place that are able to function in lieu of human oversight, such as using business intelligence analytics or automated results from quality testing equipment. The requirements of good governance are not related to strong or weak economies, they are a constant reminder that organizations and people both need to do the right thing.

The August 1, 2010 edition of the *Wall Street Journal Sunday* as printed in the *South Florida Sun Sentinel* newspaper contained results from a Deloitte LLP survey with regard to the diminishing state of trust and ethics during the current economic crisis. A reported 34 percent of the 754 employees surveyed intended to seek new employment as the economy got better. The reasons why employees were looking to leave their current employers included a loss of trust (48 percent), a lack of communication transparency (46 percent), and unethical or otherwise unfair treatment (40 percent). An employee who believed he or she is being treated unfairly or unethically may lash out at the organization and commit fraudulent acts against it.

Employee turnover can be highly disruptive – and expensive – to an organization's supply chain operations. Acquisition costs (e.g. staffing and recruitment company expenses and/or internal HR department resources), training curves until new hires get up to speed and are fully productive, loss of knowledge and skills (which can erode innovation and competitiveness) when experienced workers leave, the time required for supply chain partner (customer and supplier) relationship building and a partner's perception of the turnover, and the effects on employee morale are all ramifications of high turnover.

Organizational management should therefore not be fooled into believing that an improved economy will set everything right. Personnel must be carefully managed through tough economic times in order to reduce turnover and retain talent, enabling the organization to be well positioned to take advantage of the better business environment when the economy starts to climb out of the doldrums.

18

Trends in Ethics and Integrity

After all is said and done – after the policies and procedures have been written, after the technology to prevent penetration hacking has been implemented, after the analytical programs have traced through transactional data for anomalies – there is still one constant factor that is pervasive throughout all supply chains: People. As highlighted in Cressey's Fraud Triangle in examining the rationalization aspect of why fraud is committed, we can be our own greatest allies in the fight against fraud, though sometimes we might be our own worst enemies in how we approach the battle. As we lay out our strategies and tactics, we should not forget that people are involved and are the culprits in illicit behavior. Unfortunately, organizations receive their employees late in the stages of development – it is tough to be completely shielded from news media stories of corporate crimes, the unfaithfulness of movie stars, gangster behavior by music moguls, politicians playing political games, and cheaters whose actions vault them to the status of celebrity. Sure, we have seen white collar C-level executives in boardrooms become white collar criminals behind bars, but there are also plenty who get away with it time and time again by exploiting connections and loopholes that should have been closed a long time ago. We are suffering from a breakdown in values, morals, and principles that are wreaking havoc on not just the well-being of organizations but of society in general to the point where it seems to be the rule rather than the exception to it. As we notice our own personal interactions de-evolve, we might be aware of the intense pressure put upon us to turn the other cheek or look the other way. But, as both pressure and rationalization build, how long is it before the right opportunity presents itself and we find fraud to be not just acceptable but a necessity for some semblance of normal daily functioning?

Ethics Evaporating

The results of a 2005 survey by Deloitte and Junior Achievement reveal that the percentage of teens surveyed who believe unethical behavior is needed to get ahead dropped from 33 percent in 2003 to 22 percent in 2005. While this trend was good news, the fifth annual release of the survey in 2007 stated that nearly 40 percent of teens believed that to get ahead in school, lying, cheating, plagiarism, and even violence were necessary. The 2007 report concluded that if past behavior is any indication of future actions, the business community cannot ignore shifting lines of ethics and the potential consequences.

Since this study was undertaken, it does not appear that things have gotten any better. In 2009 a study was performed by the Josephson Institute of Ethics that surveyed 7,000 people across various age groups and asked the question: "Is lying and cheating necessary to succeed?" A total of 51 percent of responding teenagers age 17 and older replied that lying and cheating is indeed necessary to succeed versus only 10 percent of responding adults age 50 and older. While we might breathe a sigh of relief knowing that we are getting more ethical as we age, the fact is that, based on these studies, a significant majority of our workforce – starting at a rather early age – believes it not just OK but in fact necessary to compromise ethics in order to achieve success. The timeliness – 2009 – of the study cannot be overlooked, however, because baby boomers grew up in a simply different – though technologically emerging – world than that of the teenage generation represented in the study.

This trend is further supported by a University of Michigan study whose results were presented in May 2010 at the annual meeting of the Association for Psychological Science. As reported in the May 31, 2010 edition of the *South Florida Sun Sentinel* newspaper, the research analyzed the level of empathy (feelings of concern or sympathy for others) across 14,000 college students over the past 30 years. The results showed a large drop in level of empathy right after the year 2000, noting that college students are 40 percent less empathetic than their counterparts 20 or 30 years earlier.

The study researchers noted that college students in the study – sometimes known as "Generation Me" (generally considered to be anyone born in the 1970s, 1980s, or 1990s) – are viewed collectively as being "one of the most self-centered, narcissistic, competitive, confident and individualistic [groups] in recent history," and then continued by stating that "it's not surprising that this growing emphasis on the self is accompanied by a corresponding devaluation of others."

The study authors noted that the reason for this might be the very increased levels of online social networking versus face-to-face social interaction. Another area of blame stated by the researchers was the overwhelming need to enhance one's résumé to remain competitive. The belief that expressing empathy is not beneficial to establishing a successful career path was also noted, explaining that people in highly competitive situations do not feel that taking the time to listen to another person who needs some sympathy is advantageous to achieving their goal.

In the US in particular, across a variety of industries, businesses are lamenting the downturn in university students deciding on achieving degrees in math, engineering, and science curriculums. Universities complain that their new students are not prepared in mathematics and reading comprehension, and fault the secondary education system, which generally comprises junior and senior high school in the US. In March 2006 at an event at Fairleigh Dickenson University, the then-US Secretary of Education stated: "We know that 90 percent of the fastest-growing jobs require postsecondary education, but less than half of our students graduate from high school ready for college level math and science." In January 2008 the American Association of Colleges and Universities released the results of a survey it had commissioned that took place between November 8, 2007 and December 12, 2008, which involved 301 participating companies each with over 25 employees. The survey was entitled "How Should Colleges Assess and Improve Student Learning?" and reported that businesses found that college/university graduates were ready for entry-level roles but were not ready for advancement or promotion.

This is a problem in tough economic situations because organizations have little – if any – funds available for educating employees and need to rely more on on-the-job-training. Further, organizations looking to replace higher-paid skilled employees with lower-wage starters are likely to discover knowledge and performance gaps that could restrict if not actually harm the organization. But the problem is perhaps a little more acute in the present day due to the characteristics of Generation Y (those born between 1980 and 1995 according to a *60 Minutes* news piece). Also known as "Millennials," aside from being technically very savvy, they are also known to have a short-term focus, require instant gratification, and be oriented toward the results and not the process. This does not seem to bode particularly well for public companies (e.g. those that must comply with regulations or laws such as SOX) where the process matters as much as – if not perhaps more than – the results. The diametrically opposed positions of Millennial employees with those of the organization's standard operating procedures might cause enough frustration whereby a Generation Y employee rationalizes that to get to the results faster, the defined procedures need to be bypassed. (And the Generation Y employee might reason that achieving faster results will appeal to his or her manager and put him or her on a fast – or faster – track for promotion, including higher wages.) If the opportunity is there and the pressure to perform in order to get promoted is enough, the "perfect triangle" now exists where fraud could be the result due to (flawed) rationalization.

As reported by the Associated Press on September 30, 2009, the Master of Business Administration (MBA) program at the Skolkovo School of Management in Moscow will be teaching students how to handle issues such as bribery, bureaucracy, and burdensome laws. The purpose is to prepare graduates to handle the reality of corrupt business practices worldwide, including dealing with global emerging economies. The Dean of the School considers institutions like Harvard to be "a business school of the past" and that "a business school of the future has got to be different." Guest lecturers up for consideration at the time of the article were to include police agency officials and possibly organized crime figures to discuss management challenges.

Along with a better focus on reading, writing, and arithmetic, it seems that there should be an equally strong focus on educating our youth about the differences between right and wrong behavior. Unfortunately, when it comes to role models in the public eye, such as those involved in politics, business, sports, and music, it is the bad behavior that most often is given the spotlight and is emulated – if not also admired – by impressionable youth. Ethics educators will find it difficult to convince results-oriented students that the ends just do not justify the means. The spoils of bad behavior are a strong enticement to compromise integrity for immediate satisfaction.

Engaging Employees

The results of a survey reported in the June 15, 2007 edition of *CFO* magazine, where employees responded as to what actions companies could take to promote ethical behavior, are noted in Figure 18.1.

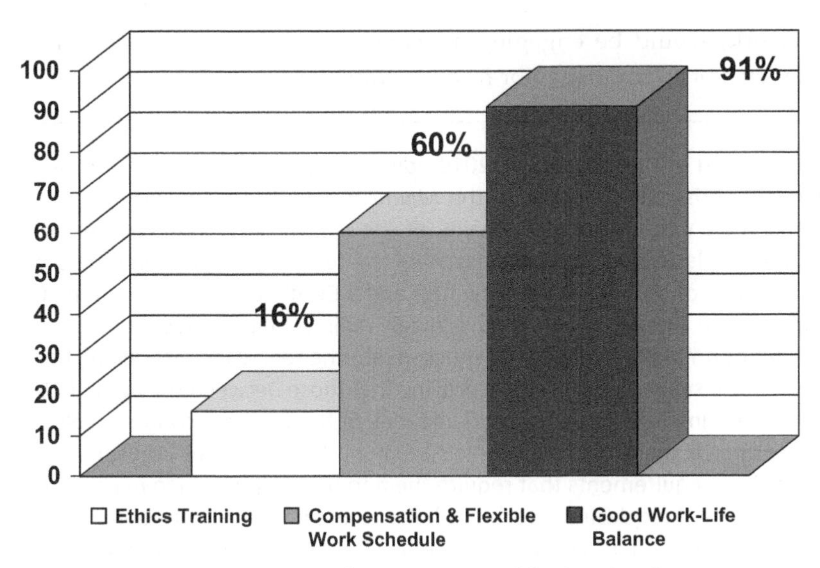

Figure 18.1 How Can Companies Promote Ethical Behavior?

> According to the University of Virginia psychologist Shigehiro Oishi: "Income disparity has grown a lot in the United States, especially since the 1980s. With that, we've seen a marked drop in life satisfaction and happiness. Certainly any [tax] policies that try to reduce income inequality would seem to be good if we care about the average happiness of Americans." Without delving into the dangerous discussion of the US income tax structure's fairness and tax breaks, we can at least take Oishi's statement and apply it to the workplace: Fair compensation leads to more satisfaction, and with more satisfaction can come greater ethical behavior.

Wait ... could it be so simple? What the results of this survey would seem to indicate is that if organizations enabled employees to have a better balance between their personal and professional lives, the employees would act in a more ethical manner. This survey was reported in mid-2007 before the worldwide financial meltdown really peaked. Let's jump to the more recent 2009–2010 timeframe: Imagine what the overworked and highly stressed employees who recognize that they are fortunate enough to have a job but are concerned about it from day to day are feeling right about now in terms of what they believe to be ethical behavior toward their employer. While I am not suggesting that promoting a better work/life balance can replace the need for technology solutions, I do think that an organization that promotes such a *tone* (recall the control environment aspect from the COSO/SOX compliance framework) would be engaging a relatively low-cost means of achieving a higher level of ethical behavior from its employees.

> As reported in the August 17, 2010 edition of the *South Florida Sun Sentinel* by Cindy Krischer Goodman of the *Miami Herald*, the recession has snapped Generation Y (back) into the reality of the current stressful economic times. As the article tells us, the business owners interviewed – including the owners or partners of both an advertising firm and a Certified Public Accounting (CPA) office – are finding it easier to manage their young employees these days. Generation Y staffers have been shocked into the real world after being handed pink slips instead of promotions, proving that those between the ages of 18 and 30 are not immune from the stark realities of the current economic doldrums. According to the article, Millennials who still have jobs are adjusting to new workplace requirements that require them to be available when needed, which means that their desired work/life balance might have to tip toward "work" more often than "life" at the present time. "This is the generation that dreamed they wanted to be CEO of a public company but didn't have an idea of what to do to get there," stated the partner of the CPA firm. The article highlighted that, according to the results released in February 2010 of a Pew Research Center study, approximately 37 percent of 18–29 year olds have been unemployed or under-employed during the current recession, which is the highest proportion among this age group in over 30 years.
>
> In one way this adjustment in attitude was needed, because Millennials still desire and deserve career development, but they have to realize that some things take time, come with experience, and require going through a process before achieving the results. Does this mean that – and according to the ethics versus work/life balance statistics above – we can expect Millennials to lie and cheat their way through their careers as they look to retain a better work/life balance and circumvent any fast-track career process? Not according to the article, which mentions that Generation Y will take this time to re-evaluate things and take advantage of learning opportunities, preparing for when the economy rebounds.

Based on the statistics reviewed so far, organizations that take steps to more closely engage their employees, e.g. by offering a better work/life balance and career advancement training opportunities, are likely to find their incidents of disruptive behavior declining. Figure 18.2 highlights the results of a study performed by the *Gallup Management Journal* and reported in its October 12, 2006 edition. The title of the study is "Engaged Employees Inspire Company Innovation."

The three types of employees are basically defined as follows:

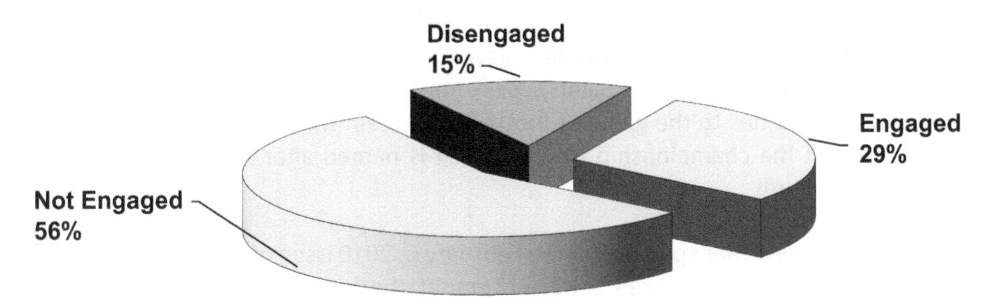

Figure 18.2 Employee Reliability

- Engaged: Employees who are innovators and moving the company forward.
- Not engaged: Employees who might be considered "warm bodies;" they do their jobs without passion and while they do not contribute to growth, they also do not normally cause disruptions.
- Disengaged: Employees who hate their jobs and/or workplace so much that they act out their anger and look to thwart the progress made by engaged employees.

Based on the survey statistics, there is very nearly a 2:1 ratio of engaged employees to disengaged employees. I would liken this scenario to taking one step backward for every two steps forward. With the majority of the employees on the sidelines, it would be difficult for an organization to make significant headway with such an imbalance in employee engagement. Yet if we could engage more of the not engaged employees and possibly even sway more of the disengaged employees to be engaged or at least not engaged, our organization would face less disruptive – and therefore less fraudulent – behavior. Peer pressure is powerful and the more people involved in committing fraud, the greater the risk of the fraud being exposed. Engaging employees as important, integral parts of the organization makes sense as a means of reducing fraud. This all might be counter-intuitive to the methods that some organizations might utilize to gain what they consider loyalty from their employees – fear of job loss. Being afraid of losing one's job is a considerable amount of pressure, and pressure that is exerted by the organization upon the employee. Organizations need to take a step back and re-evaluate the relationships they have with their employees, with a nod toward the fact that doing so should not only enable growth and stability but should also reduce wasteful costs and disruptive behavior such as fraud.

Prior to Super Bowl XLIV which was played in South Florida, the quarterbacks coach of the New Orleans Saints, Joe Lombardi, was interviewed about his unique family ties to the game of football. The team that wins the Super Bowl is awarded the championship trophy, which is named after Joe's grandfather, Vince Lombardi.

In the interview that appeared in the February 6, 2010 issue of the *South Florida Sun Sentinel*, Joe was asked about his grandfather's famous quote: "Winning isn't everything, it's the only thing," stating first that he did not agree with the saying, but then qualifying his answer. "I think his point was that certainly your goal is to win, but it's the process that you take in order to get there, the will to win, the will to prepare, doing everything in your control in order to put yourself in position to win at anything. And I think that's what his point was, more so than win at all costs even if you've got to cheat or cut corners." Well said Joe – I think your grandfather would be very proud. It is a great lesson for everyone, from Millennial to middle-aged.

Trust but Verify

The acquisition of employees is a critical component of the supply chain because without employees no supply chain could function and most would probably be unnecessary. As is standard, prospective employees submit a résumé or curriculum vitae containing work experience, skills, qualifications, certifications, education, etc. Depending on the job role, it is critical to establish the veracity of a person's background. Lying is a bad way to start a relationship. Sometimes people lie on their résumé due to desperation, such as if they have been out of work for an extended period of time and are facing financial hardships, but it is still a fraudulent act. This may be a reflection on the person's fortitude when faced with other challenges, such as those in the workplace – will the person react to hardships on the job with illicit behavior too?

As reported in the February 5, 2010 issue of the *South Florida Sun Sentinel* and attributed to McClatchy/Tribune News, the University of California (UC) has created a four-person group of auditors that randomly selects and verifies the information on student applications. Due to the highly competitive nature of UC's acceptance program, the education system looks to screen out applications that appear too good to be true. American universities can ask for verification of grades from high schools, community colleges, and other universities, as well as officially administered test scores. However, other information on the application is rarely if ever validated, being left up to the honor system. Though UC is only auditing one percent of the 134,000 applicants at the time of the article, those who have embellished their accomplishments will likely find their applications rejected. UC hopes that the fear of being audited will be enough to keep most applicants honest about their achievements.

Falsified education and certification credentials should be easy to expose simply by contacting the education institution or certifying association. It is more common for employers to perform credit checks on prospective employees to ascertain whether a person is financially responsible and if they are under a heavy debt load, the reasoning that people with bad credit scores might be more prone to steal from the organization to relieve their financial distress.

The background check for criminal activity is a common pre-employment screen. However, the Association of Certified Fraud Examiners (ACFE) – in both its *2008* and *2010 Report to the Nations* – reported that 86 percent of fraudsters are first-time offenders, meaning that the person had no prior charge or conviction of a fraud-related activity. The ACFE concluded that the effectiveness of background checks is "probably limited" as an indication of a person's potential to perpetrate fraud.

The problem with background checks might be twofold: One is that we get what we pay for and another is that it is not yet an exact science. Background checks are typically performed by third parties who purchase information from – in the US – states, counties, and cities. When qualifying a background checking company, it is important to ask how recent the data is that is being used: Is it one month old, one year old, or older? Cities, counties, and states earn revenue for selling this information, but this means that the independent companies offering background check services must invest in keeping the data current. A prospective employee may have committed an act recently that would disqualify him or her from employment, but if the background check service company does not keep its data up to date, it might not show up in the prospective employee's profile.

Another issue – at least in the US – is that the most information about a person is at the county level, not the federal, state, or city levels. County records include births, deaths, marriages, divorces, and property ownership. In the state of Florida there are 67 counties, some more technologically sophisticated than others. Third-party background check companies would not be able to acquire much information from small counties that have not converted or implemented electronic record-keeping instead of paper processing. Therefore, information about a person who lived in a particular county might not be included in a background check because the data was not current or was simply not available in electronic form for purchase. Understanding the limitations that a background check company faces is important before using its services and making important hiring decisions based upon the information reported back.

However, even with most information about a person held at the county level, the location of criminal records might depend on whether the person resided at the time in a geographic area served by a city police force or a county sheriff's agency. According to the FBI's website, the FBI annually compiles (and has since 1930) the Uniform Crime Report (UCR), which is a "voluntary city, university and college, county, state, tribal, and federal law enforcement program that provides a nationwide view of crime based on the submission of statistics by law enforcement agencies throughout the country." The statistics for the UCR come from "nearly 17,000 law enforcement agencies across the United States." The FBI's 2008 *Crime in the United States* report – one of the reports created under the UCR structure – was generated with the input of 14,169 police agencies in the US. Performing a comprehensive background check on a person could require verifying information with several police agencies depending on where that person lived because of all the agency independence and without a central data warehouse for all individual's criminal records nationwide.

For most employees, the low-cost background check provides some semblance of security in the decision to hire a person. For people applying to executive management or other high-level positions, a more comprehensive background check should be performed, and this might require visits to local courthouses or police stations to acquire copies of records to provide support for information found – or not included on – a person's résumé.

Trust and Verify

Keeping with this theme but jumping topic a little, it is fine to have a trusting relationship, but sometimes that trust is because each party knows the other is verifying. Take the examples of tainted product frauds: Had both the customer and supplier been performing quality assurance checks – and making the results known to the other – I think it is likely that the tainted products would not have reached consumers. Sharing detailed data – sales information, forecasts, shortages, capacity constraints, logistics logjams, quality control data, etc. – is very difficult because it requires both trading partners to really open up and put on the table sensitive information about their respective organizations. However this intimacy results in better collaboration and the ability to address small issues before they become big problems.

Just because you can outsource a process like manufacturing or a service like documentation review does not mean that you can outsource the responsibility – that accountability rests solely on the shoulders of the customer organization. Reducing fraud requires that both the customer and supplier organization work closely together in what should be a trusted relationship. The perception of detection is a power tool in the fight against fraud. However, the reality is that the detective work must be sufficiently performed against a large enough sample quantity to catch anomalies and stop the problem from advancing further through the supply chain, and sometimes we have to accept that this requires 100 percent inspection even if that means at a higher cost of operations. Reducing fraud results in improved supply chain performance and is simply the right thing to do. Remember that the life an organization saves through diligence and dedication might be that of one of its own employees or an employee's family members or friends.

In the February 2011 issue of *CSO* magazine, in an article entitled "Helping Employees Really 'Get' Company Policy," Michael Santarcangelo of Security Catalyst defines what he calls the Human Paradox Gap as "the distance between making a decision and feeling its consequences." He states that: "When people are disconnected from the consequences of their actions, they do not take responsibility and are not held accountable. More technology and more training misses the point. The underlying challenge is the disconnection." He believes that instead of focusing employees on security and risk, the focus should be on responsibility, consequences, and accountability. Global supply chains have removed people from close proximity to the effects of their decisions. Perhaps this is an interesting "fourth corner" of Cressey's Fraud Triangle, in that relative distance may be a factor in a person's decision-making process when considering the perpetration of fraud.

19

Supply Chain Information Integrity

To reduce incidences of supply chain fraud, organizations need to focus on their employees. In general, but especially in tough economic times, employees may find themselves overworked, underpaid, undertrained, and underappreciated. It is especially important that during the difficult times organizations do not take for granted the human element. The March 2009 issue of *Inbound Logistics* magazine included the results of a survey that listed the top 12 corporate ethics and compliance concerns among the respondent companies. The results are listed below:

1. Anti-bribery requirements.

2. Conflicts of interest and gifts.

3. Antitrust contact with competitors.

4. Mutual respect.

5. Records management.

6. Product safety and liability.

7. Privacy.

8. Proper use of computers.

9. Export controls.

10. Careful communication.

11. Information security.

12. Financial integrity.

What I personally found a little disturbing was that product safety and liability ranked in the middle at number six. As a consumer of goods, I would like a little more assurance that the foods I am ingesting and the products I am using are safe. Given the tainted product frauds that have plagued the planet in the past few years, my confidence is somewhat shaken when I go to the store but I have few, if any, practical options available to me.

It should also be noted that information security and financial integrity ranked at the bottom of the list. I do not know the breakdown of public versus private companies in the survey, but I think that – especially with SOX now being well established – public companies would have a need to focus on financial integrity and it should rank a lot higher. Recent guidelines from the US Securities and Exchange Commission require public companies to report security breaches to investors. Individual states are enacting their own versions of data breach notification laws. We have already seen how reducing fraud in the supply chain operations process and systems helps to improve an organization's financial integrity.

The fact that information security ranked so low is also a concern because of the sensitive nature of the data used to drive supply chains. As reported in the November 30, 2009 issue of *Information Week* magazine, in a Cyber-Ark survey of 600 financial industry employees in New York and London, 41 percent admitted to having taken sensitive data from one job to the next, even though 85 percent of the respondents admitted that the downloading of such information – let alone walking away with it – from their employer is illegal. The data of choice was primarily customer and contact lists. In other industries such as manufacturing, component information in the bill of materials could reveal secret ingredients. Prices paid for raw materials and the vendors from which they are purchased also represent competitive information. This survey reported that even a hint of job loss caused some employees to spring into action (39 percent) and take sensitive data, and that 48 percent of respondents stated that they would take company information with them if they were fired. Thirteen percent said they would take network log-in information (user identifier and password) for the purposes of accessing their former employer's network after their employment.

These statistics seem to agree with those reported in the April 2009 issue of *CSO* magazine in an article entitled "Laid-off Workers as Data Thieves." In this survey 1,000 former employees were polled within the prior 12-month period and the results are quite revealing:

- 53 percent admitted to stealing confidential data.
- 79 percent took data without an employer's permission.
- 82 percent said employers did not perform audits prior to dismissal.
- 24 percent had access to computer systems after dismissal.

With information security ranked near the bottom of corporate ethics and compliance concerns, is it really any wonder that so much data theft is taking place? It should also be noted that all of this data theft is being perpetrated not via malicious network hacks by unknown bandits but by the organization's very own existing and former employees who have full access to the computer systems. Whether theft of data or theft of monies via fraudulent transactions (as highlighted in some of the examples in this book), organizations seem to face more threats from their own employees than from outside forces, yet are not protecting themselves appropriately. And most of these frauds are happening within the protected network infrastructure by people (employees) on the inside of the organization.

In the March 30, 2009 issue of *Information Week* magazine, the results of a survey of 400 respondents regarding senior management top security priorities was published. Notable statistics from the survey are in listed below:

- 41 percent – Regulatory compliance.
- 35 percent – Protecting data from outsiders and hackers.
- 32 percent – Keeping security costs to a minimum.
- 27 percent – Understand and manage security risk.
- 23 percent – Enable employee access to useful data.
- 18 percent – Protect data from unauthorized emp. access.

- 15 percent – Project an image of security to outside world.

One important feature of this survey is that 35 percent of senior managers focused on protecting data from outside attacks and only 18 percent prioritized protecting data from unauthorized employee access. Given that 32 percent of respondents were looking to keep security costs to a minimum, these statistics seem to support a lack of focus on where much of the data loss is really happening. Granted, securing data from employees whose job roles are to manage (IT staff) and maintain (end users) this information is a challenge and can be something of an oxymoron. How does an organization restrict access to something that – perhaps simply by its own nature – must be accessed by everyone? Shutting down hardware access points like USB drives on workstation computers, disabling mass download access in ERP systems (and the like) via the software's user rights and roles policies, restricting IT staff access to data to those who absolutely need it, and ensuring written and signed policies are in place are all steps to begin to control this problem. Nothing will happen until there is recognition by executive management that this problem actually exists. Unfortunately, either via their own perceptions or due to a reliance on what they are being told, senior managers do not seem to be focused on this area.

In developing their "2009 Global Risk Management Survey," Aon reportedly received input from 551 organizations from around the world, including government agencies, private enterprises, and public companies from over 40 countries between October and November 2008. The top ten global risks – in order of their priority from top to bottom – are as follows:

1. Economic slowdown.
2. Regulatory/legislative changes.
3. Business interruption.
4. Increasing competition.
5. Commodity price risk.
6. Damage to reputation.
7. Cash flow/liquidity risk.

8. Distribution or supply chain failure.

9. Third-party liability.

10. Failure to attract or retain top talent.

Given the timing of the survey, it is not surprising that economic slowdown was the number one concern. It should also be noted that supply chain failure is ranked quite low, even though it is linked to higher-ranking risks such as business interruption and damage to reputation (e.g. from tainted products). One would think that supply chain failure would therefore be ranked higher. What is missing on the list is anything to do with security of the organization's information assets. Based on the various statistics presented here from different sources, it would seem that there is a quite a large understanding divide or perception gap between what is really going on in the workplace in terms of supply chain frauds such as information theft and what executive or senior management believes is really happening: That their information is less secure than they realize.

This all seems to completely contradict the prior statistics with regard to employee training to improve ethical behavior, but the world has changed since the above survey was released in 2007, and during that time there were two surveys reporting a decline in the belief in ethical behavior by teenagers. As the worldwide economic situation gets worse, it is bringing out equally worse behavior – perhaps out of desperation – in what seems to be an increasing number of people, including the next generation who will take over the supply chains we all rely on so heavily today. Without members of senior management focusing on the growing internal threat within their organizations, the risk of information theft and the negative consequences it can bring will continue to be a risk hiding in the shadows.

In its June 2011 issue, *CSO* magazine published its "2011 State of the CSO" article subtitled "The Rise of Risk Management," which analyzed the results of responses from 229 security professionals covering industries – in descending order of response rate percentage – such as financial services, government and non-profit, manufacturing, high technology, telecommunications and utilities, and health care. Respondents reported their primary job roles as being in the areas of information security, privacy, fraud prevention, investigations and audits, and personnel security.

One of the statistics presented is called "Winds of Change" and asks the question: "Which of the following trends will have the most profound effect on the role of security professional?" Technology as a service and technology that allows people to be connected 24/7 garnered responses of 21 percent and 27 percent respectively. I see these as being somewhat related, given that technologies like cloud computing and mobile devices allow round-the-clock access to applications where data is stored.

Next-generation workers coming into the workforce and the use of social medial were concerns for 21 percent and 16 percent for the respondents respectively. Again I see these as being related, because the next generation of workers – the Millenials/Generation Y – have grown up not just with but have become really, it seems, dependent upon social media technology and may be the largest user demographic of social media technology. What causes security professionals to be so concerned about the next generation of employee? The *CSO* survey does not provide the reasons for the respondents' answers, though issues such as a lack of integrity and ethics could be factors. Competitive intelligence professionals are able to gather a wealth of information about an organization from news articles, classified advertising, blog posts and now social media updates, and this provides an additional challenge for security professionals to control.

PART IV
Supply Chain Vendor Compliance

20

Collaboration versus Conspiracy

In the May 2009 edition of *World Trade* magazine, I read that three more global air carrier firms were found guilty of price fixing by the US Department of Justice. Based on the article, it appears that over the course of six years, a total (at the time) of 21 rival companies conspired to set service pricing. As reported in the news on July 30, 2010, Delta Airlines will pay a fine of $38 million to settle criminal accusations regarding the cargo unit of Northwest Airlines, a subsidiary of Delta as of 2008, having met with other airlines from 2004 to 2006 to conspire to fix prices. During the period in question, Northwest Airlines was the only US airline with a dedicated fleet of air freighters: Air carriers typically carry cargo mixed with luggage in commercial flights. The total fines exceed $1.5 billion and three corporate executives (thus far) have been sent to jail. In fixing their prices across the industry, these companies were robbing their customers of competitive leverage in choosing air cargo services. Price fixing is fraud and is illegal in the US, where the Sherman Anti-Trust Act and the RICO (Racketeer Influenced Corrupt Organizations) Act can both be used to prosecute organizations that conspire to rob their customers of competitive goods and services.

At the same fraud conference in March 2011 where I heard Mr Owens speak, I had the opportunity to also hear from Mark Whitacre and his involvement in the price-fixing scandal at his former employer ADM. Mr Whitacre described how ADM and 11 of its competitors conspired to fix prices whereby each participating company benefited by approximately $1 billion per year. Though Mr Whitacre assisted the FBI by wearing a wire to record meetings for three years, he nonetheless served over eight years in federal prison for his role in the scandal, which included price fixing, tax fraud, and money laundering. His involvement in this fraud was the basis for the movie *The Informant*.

What is particularly interesting about this case is that it was Mr Whitacre's wife who, upon learning of what her husband was doing from him, forced him to reveal the fraud to the FBI, which began his three-year involvement in the investigation. Mr Whitacre stated that while he was in prison, the affected consumer product companies that were previously paying inflated prices for ingredients sought out his wife – the real whistleblower – and provided financial support for housing and education during Mr Whitacre's incarceration as a way of saying "thank you" because of all the money they were saving by paying true competitive prices instead of artificially fixed prices.

When service companies conspire to set prices, there may not be as much pressure on the part of the conspirators to perform for competitive purposes. So what if performance slips a little? It is not necessarily likely – on the surface – that any one company's customers will jump ship for better pricing, especially if overall performance is still within tolerable limits. But there are ramifications to the illegal act of conspiracy that ripple through the supply chains of the conspirator organizations' customers: As performance slips, disruptions to supply chains grow and other costs can increase. Modern supply chains have a greater reliance on just-in-time (or nearly just-in-time) replenishment, whether it is the retail store shelf or the distribution center bin. The impact of just a few percentage points slippage in performance can result in empty shelves, equating to lost retail sales or idle manufacturing shop floors depending upon the point in the supply chain. The greater the reliance normalcy, the lower the resistance to disturbances that interrupt the smooth flow of things. Essentially requiring customers to expedite more goods and hold higher levels of inventory, this fraud penalizes the customers by forcing them to incur greater operating costs. The cost of goods sold increases and unless the organization is willing to absorb the expenses – which will have a negative impact on the financial bottom line – the organization is forced to pass along costs increases to consumers as higher prices. Thus, we consumers can be left footing the bill for these kinds of industry conspiracies!

When individuals use their job role to perpetrate fraud, it is considered *occupational* fraud. When executive management perpetrates fraud – essentially committed through the organization itself – it is considered *organizational* fraud. Though the term typically refers to fraud against an industry or business in general, I think it would not be incorrect to label fraud perpetrated when competitors conspire as *industrial* fraud.

What is not illegal is for organizations – including competitors within an industry – to come together to set standards and guidelines for business practices and supply chain relationships. One of the leading – if not pioneering

– industries in the establishment of such standards and guidelines are the US retailers. A visit to the website (www.vics.org) of the Voluntary Interindustry Commerce Solutions Association (VICS), the retail industry's trade association, reveals standards and guidelines for the bill of lading and how merchandise should be shipped in preparation for display on the retail floor with minimal intervention by the store staff. Known as the Floor-Ready Guidelines, even the types of hanger to use are specified. EDI is the standard method of retailer–supplier electronic communication of business transactions (purchase orders, shipping notices, invoices, etc.), and at least there is consistency, even if not complete uniformity in the format of the EDI transactions and those of the item, carton, and pallet barcode labels, a disparity that causes its own unique set of challenges for suppliers who sell to multiple retailers.

The rules which govern how suppliers interact with their customers are called *vendor compliance* requirements. Vendor compliance documentation is prepared by the customer (e.g. the retailer) and brings standardization to the buying party's relationship with its supplier base. (This is predicated on the buyer being able to leverage against its suppliers. Typically size is the determining factor whereby the customer is simply a much larger organization than any of its suppliers. However, in retail the suppliers recognize that retailers have the distribution networks and stores that they lack to convey their goods to consumers, and there are suppliers who are larger in terms of annual sales volume than some of the retailers they sell to.) Vendor compliance documentation typically includes data mapping guidelines, routing and logistics information, paper documentation (e.g. packing list) formats and requirements, barcode label (item, carton, and pallet) diagrams, key contact information, vendor scorecard information, and chargeback amounts for non-compliance.

Organizations in all industries have not only the desire but generally also have the need to operate as cost-effectively as possible. This is no great revelation. The need for retailers to reduce costs through supply chain optimization and efficiency is considerable. "Black Friday" is the day after the last Thursday in November Thanksgiving holiday in the US, when retailers traditionally hope to begin turning a profit with the end-of-year holiday shopping spree after running in the red (at a loss) for the year to date. Any ability to trim operating costs is a necessity, especially in tough economic times, particularly for a business model that basically operates in the negative 11 out of 12 months of the year. This is not necessarily true for all retailers, especially those with more diversified business models, though few would dispute Black Friday's impact to retailer profitability as the signal to the start of the holiday shopping season. Through standardization of the touch-points with its supplier base – and a typical US retailer can have from a few thousand to several thousand suppliers – the retailer gains efficiency through conformity.

Adherence to the rules of the game are enforced with vendor compliance *chargebacks*: Financial penalties taken against the invoice remittance (deducted from the supplier's invoice payment) for failure to comply with the established vendor compliance guidelines. (In the pharmaceuticals industry a *chargeback* is a rebate on contract pricing, not a penalty for non-compliance with business-to-business guidelines. Nonetheless, the pharmaceutical industry, and notably the large distributors, does financially penalize its suppliers – pharmaceutical product manufacturers – for non-compliance with its supply chain requirements.) Chargeback deduction reasons and the penalty amounts are (or at least should be) listed in the vendor compliance documentation.

Common chargeback reasons include non-compliance with:

- item ticketing requirements;
- unreadable, improperly formatted, or missing carton and pallet barcode labels;
- excess or unapproved packing material;
- missing or improper documentation (e.g. packing list, bill of lading);
- incorrect hangers;
- incorrect EDI transaction data or improper record or file structure;
- late shipments;
- late EDI transactions;
- incorrect shipments (e.g. the wrong item was shipped);
- short shipments (less than 100 percent of a requested item was shipped).

The vendor's scorecard is a means of tallying and grading a supplier's performance and adherence to vendor compliance requirements. The scorecard should be based on key performance indicators against defined metrics. For example, a common metric is late shipments: Whether the shipment arrived on time or later than the required delivery date. The key performance indicator is set by the customer, e.g. the customer requires at least 95 percent of all shipments to arrive on time. Other common key performance indicators seem fairly straightforward, e.g. either a barcode label exists on the carton or it does not. The readability of a barcode – whether on a carton or on a product – is another key performance indicator common in supply chain vendor compliance.

Being at the forefront of supply chain optimization through the collaborative efforts of some of the oldest and most well-known retailer brands in its history, the US retail industry serves as an example to not only newer emerging retailers – whether traditional brick-and-mortar or online retailers – but also to other verticals as they look toward the experience of US retailers when developing their own industry's supply chain standards and guidelines. The US retail industry is well known in certain circles for penalizing its suppliers that fail to conform to standards. The complexity is exacerbated because the industry's standards have grown such that they are something of a "non-standard standard." For example, the carton barcode label format can vary from one retailer to another, yet each one is compliant within the overall industry guidelines. Data formats in the same EDI transaction (e.g. the purchase order, invoice, or advance ship notice) between multiple retailers can vary and yet all can be valid within the broad standard that EDI has become. The US retail industry may be more the case of "Do as I say not as I do" with regard to the commitment to truly defined industry standards. Is it legal and ethical to penalize suppliers for non-performance against shifting, inconsistent requirements and without the ability to inspect what they are penalized for?

The Cost of the Corrections

Vendor compliance is an expense to suppliers and one that suppliers rarely embrace wholeheartedly. Since my first introduction to retail vendor compliance in 1993, I have only heard suppliers bemoan the fact that they need to incur costs to help their customers operate more efficiently, taking a "what's in it for me?" position. The truth is that – more and more – the closer a supplier can collaborate and be a less disruptive partner, the better the customer–supplier relationship and, hopefully, this translates into more sales that then cover the cost of compliance and still leave a tidy amount for profits. And many times the supplier can utilize the technologies being implemented to improve its own operational performance to increase efficiency and accuracy and support its own fraud detection and reduction programs.

One of the most straightforward gains for a supplier comes from the requirement to barcode label items and shipping cartons. Item barcodes can be used to verify the pick-and-pack process as well as for more efficient and accurate inventory counting. The unique carton label barcode identifiers can be used to validate shipment staging when many cartons are involved and to verify shipment completeness, helping to ensure that no cartons are left behind. This is an example of how a project to improve operations – albeit based on a customer requirement – can be used to also help detect and reduce fraudulent behavior. As I have previously mentioned, the use of automatic identification technology such as barcode label scanning transitions an organization from a paper-based to a paperless operation and enables the data to be programmatically audited and cross-checked for accuracy and integrity.

The introduction of chargebacks as a penalty stems from the fact that retailers – and this would be applicable to other large organizations such as automotive manufacturers, hospitality (e.g. in the food and beverage areas and in the various retail and sports shops), and pharmaceutical distributors – simply need their suppliers to comply for the sake of maximizing operational efficiency and, because they have the leverage to do so, can penalize their suppliers for disruptive behavior. In theory, chargeback amounts are supposed to be just enough to cover the cost of the corrective actions and not more. However, the gap between theory and reality can sometimes be a chasm: Chargeback amounts often have the appearance of exceeding the cost of the correction and therefore their validity and legality comes into question. The difference between the perceived costs of correction versus the believed costs of correction is a significant bone of contention between the supplier and customer, especially in retail: The strain is noticeable and the animosity palpable in my experience.

Further complicating matters is the issue of whether the retailer's staff – from office personnel to the distribution center – and technology are capable of fairly and consistently assessing a compliance violation as judge, jury, and executioner of the financial penalty as the punishment. In the US there is a basic presumption of innocence until proven guilty by the submission of evidence to the contrary, and vendor compliance programs tend to operate against this basic principle. Too often, vendor compliance chargebacks are assessed at the sole discretion of the retailer, without input from the supplier, sometimes weeks or months after the infraction date, combining to make it impossible for the supplier to refute. Because these chargebacks are happening at the retailer's corporate offices, distribution centers, and stores, the supplier does not have ready access to these locations and is subject to the sole discretion of the retailer. Incorrect chargeback reason assignment and uneven assessment of financial penalties for the same reported problem causes both the penalty amount and the determination capability of the retailer to be called into question by the supplier – it can be a daunting if not impossible task for a supplier to ascertain why a retailer assessed a chargeback one time and not another, and sometimes the same infractions are classified by the retailer differently, making problem determination and correction a costly and time-consuming challenge for the supplier.

The problem with the evolution of the chargeback is the opinion by many retail suppliers that these financial penalties have become profit centers, helping retailers to cover losses in other areas of their organization. Compounding this is not just the belief but the certainty within the retail supplier community that chargebacks are selectively enforced and waived in backroom deals between retailers and very large suppliers with desirable brand names that can draw consumers to stores. This equates to a tax break for the rich while burdening the lower and middle classes – in this case the small, medium, and even large suppliers – with increasingly stiff financial penalties for non-performance to compensate for relinquished chargebacks that might have been forthcoming from select brand-name suppliers.

At a supplier symposium I attended, even well-recognized brand suppliers with annual sales in the hundreds of millions of dollars per year – some even close to the $1 billion mark – were complaining of suffering from excessive vendor compliance chargebacks. One of the conference presenters stated rather matter-of-factly that suppliers should just engage in talks with their retailer trading partners and negotiate chargeback amounts. The speaker's statements fell on disbelieving ears as some of the suppliers had relayed their vendor compliance horror stories to the group just moments before. The fact was that the company the speaker represented made several billions of dollars in annual sales, equivalent to and even larger than the size of some of its retail trading partners, and was thus able to leverage size and brand strength to its benefit in the chargeback negotiation process. Only a very small percentage of retail suppliers would have such brand-name recognition and size to be able to leverage against their retailer customers. This speaker represented several well-known apparel brand names with high consumer loyalty. There can even be discrepancy in supplier relationships between retailers and different very large suppliers, such as apparel versus personal consumer products: Loyalty to an apparel brand may be more intense than loyalty to a brand of laundry detergent, for example, something that the retailer will be aware of and will understand in its dealings with different suppliers. The introduction of private-label goods may also have an effect on the retailer–supplier relationship, placing the brand-name supplier at more of a disadvantage for space on crowded store shelves or clothes racks.

In my experience, chargeback notification to suppliers is inaccurate, convoluted, difficult to understand, and can be weeks or months after the supposed infraction was said to have occurred. There is a lack of standard retail industry terminology or identification for chargebacks, further adding to the supplier's burden to translate why it was financially penalized. Suppliers are at the mercy of retailer's judgment to determine and assess (correctly assign the correct chargeback code and penalty amount) accurately. Uneven enforcement, pressure for profits, and a lack of sufficient controls all combine to create an environment where potentially unethical and possibly fraudulent behavior on the part of the retailer can occur. We cannot rule out that a supplier might be exerting influence in the form of gifts (also known as kickbacks or loans) to distribution center staff to look the other way on compliance violations.

As reported in the August 20, 2010 edition of the *South Florida Sun Sentinel*, police officers in the city of Sunrise, Florida can be given a written reprimand for not making enough traffic stops during the course of a year. The city denies having a quota system in place but does not refute that it expects its road patrol officers to make at least three traffic stops per day. At the time of the article, the city had 84 road patrol officers who each worked 182 days per year, equating to 45,864 traffic stops per year in total.

The city's position is that this guideline enables administrators to determine whether an officer is performing his or her duties and ensure that the officers are productive and actually working during the time they are on duty. There are instances across various police agencies in South Florida – and I am sure across the US and elsewhere – where police officers have been caught working at other jobs, sleeping, or otherwise engaged in activities not related to police work while they were on-the-clock.

The article states that, according to police union officials, officers who meet or exceed the quota – known as a "shift standard" – are more eligible for special assignments, promotions, and pay increases versus those whose performance falls under the minimum quota mark. It is the contention of many suppliers that retailer staff – especially those in the distribution centers – have similar quotas that affect bonuses and employment opportunities. These allegations further strain the relationship between suppliers and retailers as some suppliers contend they have proof of exactly this kind of quota system for chargeback assessments.

Let us remember Cressey's Fraud Triangle, in which three things have to occur to enable fraud to be triggered: pressure, opportunity, and rationalization. All organizations place pressure on their employees to perform, but retailers – with a business model that makes up significant annual profits during the last month of the year – have a somewhat unique pressure placed upon them. The opportunity to commit fraudulent behavior exists in that the retailer's own personnel and business applications – both at a distance from the supplier –

solely determine whether the supplier is in or out of compliance. Rationalization may not be enough to thwart excessive pressure and opportunity if in fact – as some suppliers believe – employees at retailer distribution centers are pressured and even incentivized to meet certain chargeback goals or face dismissal for failure to meet those same goals. The latter is an example of the "fraud or flight" syndrome described earlier where an employee – faced with job loss unless certain unattainable goals are met – can actually be forced to commit fraud by the very organization he or she works for.

A note to readers: Please remember that I am not an attorney, barrister, or any other type of law practitioner. I will be referencing and using the content of the Uniform Commercial Code (UCC) on the Cornell University Law School website (www.law.cornell.edu/ucc/ucc.table.html) as the basis for the analysis that follows, which is not meant to form a legal conclusion or provide legal advice.

Are Chargebacks Legal?

Several sections of the UCC infer that the notion of a chargeback may be allowed. Section 2-714 (Buyer's Damages for Breach in Regards to Accepted Goods) states that: "Where the buyer has accepted goods and given notification [subsection (3) of Section 2-607] he may recover as damages for any non-conformity of tender the loss resulting in the ordinary course of events from the seller's breach as determined in any manner which is reasonable." Non-conformity of tender could apply to a violation of vendor compliance guidelines. But this section further states that: "The measure of damages for breach of warranty is the difference at the time and place of acceptance between the value of the goods accepted and the value they would have had if they had been as warranted, unless special circumstances show proximate damages of a different amount." Does a violation of vendor compliance guidelines change the value of the goods being tendered? How does an incorrect hanger, missing item tag, unreadable carton label barcode, late Advance Ship Notice EDI transaction, or excess packing material alter the value of the item the supplier sold the retailer? Do these translate into an item not making it to the store and thus delaying its availability to be sold? Aren't some of these just operating costs that the retailer has to bear? And if the retailer is not happy with the performance of a supplier, should it not simply dismiss the disruptive supplier and find another to take its place? Wouldn't this be easier if not cheaper than the overhead of managing a problematic vendor?

According to Section 2-717 (Deduction of Damages from the Price) of the UCC: "The buyer on notifying the seller of his intention to do so may deduct all or any part of the damages resulting from any breach of the contract from any part of the price still due under the same contract." The implication seems to be that at the minimum the subject of chargebacks should have at least been included in the contract between the retailer and the supplier such that there was full disclosure of the potential for financial penalties for compliance violations. Hopefully the retailer – if even just upon request – would provide the list of chargebacks to the supplier to review along with the contract before any signatures were applied and the business relationship between the retail and supplier was formalized.

The term "damages" is not defined within the UCC, so its meaning is somewhat subjective. From the retailer's perspective, a chargeback could be argued as a type of damage to normal operations processing which results in incurring additional costs to correct the compliance violation. There is no apparent necessity within the UCC that damage be something physical in nature. Therefore, a retailer could argue that the administrative costs of managing supply chain disruptions are a type of damage that the retailer is legally permitted to collect, but is this allowed?

Section 2-718 (Liquidation or Limitation of Damages; Deposits) of the UCC states that: "Damages for breach by either party may be liquidated in the agreement but only at an amount which is reasonable in the light of the anticipated or actual harm caused by the breach, the difficulties of proof of loss, and the inconvenience or nonfeasibility of otherwise obtaining an adequate remedy."

This is where much of the argument between suppliers and retailers exists: What is the true cost of the correction for the disruption caused by the compliance violation? Different compliance failures will carry a different cost of correction, e.g. a barcode on a carton that cannot be scanned due to damage (e.g. wrinkling) versus a missing barcode label where more effort is required to resolve and ensure that the correct data is entered into the warehouse management system's receiving function.

What throws a proverbial "monkey wrench" into the supplier–retailer relationship is that the supplier is held responsible for compliance violations that can occur during the shipping process. The supplier is typically not able to select the freight carrier, but must adhere to the choices specified by the retailer and noted in the retailer's routing guide. The assumption that the freight carrier may not be responsible for carton damages, mutilation of barcode labels, or missing merchandise is somewhat naïve. The fact is that just because there is a contract between the retailer and the freight company, it does not mean that the freight carrier is ruled out as a suspect when investigating the causes of compliance chargebacks. However, it is the supplier that is financially penalized for violations outside its control and must bear the burden of proof against the retailer and the retailer's transportation partner(s), which rather stacks the deck further against the supplier being able to refute financial penalties, including those that did not happen while the goods were under its control.

Compliance chargeback fee calculations vary from a per-incident fee to a flat fee to a combination flat fee plus a per-incident charge, sometimes with a minimum amount to be assessed. Are the amounts charged reasonable and do they reflect the actual harm suffered by the retailer? In my experience, most suppliers would argue that the compliance chargeback calculations are in excess of the actual cost of the correction. Some vendor compliance manuals may state that the chargebacks cover both the additional costs that the retailer incurs in exception handling plus the cost of associated administrative fees. These administrative fees seem to be covered under Section 2-715 (Buyer's Incidental and Consequential Damages) of the UCC, which states that: "Incidental damages resulting from the seller's breach include expenses reasonably incurred in inspection, receipt, transportation and care and custody of goods rightfully rejected, any commercially reasonable charges, expenses or commissions in connection with effecting cover and any other reasonable expense incident to the delay or other breach." However, the cost of administering the chargeback is still in question. Is the cost covered by also charging a percentage of the shipment or order, or is such a calculation in excess of the cost of the administration? The answer probably lies in time studies, which look at the amount of human effort it takes to resolve and administer the infraction with the time-cost of the person performing the work in the distribution center and office.

Thus far, the establishment and assessment of chargebacks all seems quite legal, even if the cost of the correction is in question. Retailers have the ability to establish chargebacks and set fees that should be based on the true cost of correction and can include the cost of administering the chargebacks. This is very important because suppliers can suffer compliance violations that range from tens of thousands to hundreds of thousands – and even into the millions – of dollars per year.

We should recall that just because something is legal to do does not mean that it is also ethical to do. The city of Sunrise police chief was quoted in the article as stating that "I don't have a ticket quota and I don't have an arrest quota. It's not illegal, but it's unethical." While the question of the legality of chargebacks seems to be answered, there remains the question of the validity of the amount of the chargeback and if the chargeback programs are ethically operated by the retailer.

Are Chargebacks Valid?

At a retailer supplier conference I attended, a shoe manufacturer provided the following story to the audience in attendance.

The shoe manufacturer was suffering considerable chargebacks for damaged boxes to the tune of tens of thousands of dollars (and perhaps even into six figures – the supplier did not want to divulge the true amount) that had accumulated during the course of one year. Upon contacting the particular retailer where the majority of the problems were occurring, the supplier and retailer agreed that the supplier would be on-site at one of the retailer's distribution centers to observe the unloading of one of its shipments. Only the distribution center manager supposedly knew of the supplier's visit; none of the other distribution center personnel were aware. The supplier's shipment arrived and the distribution center staff began their routine of unloading the cartons of shoe boxes. Because the retailer's vendor compliance guidelines stipulated "floor loading" (no pallets were to be used), the cartons were all neatly stacked in the truck trailer. (Now here is where I personally find the cleverness of some people to really shine!) In order to get at the cartons at the top of the stack in the truck trailer, the distribution center personnel unloaded the cartons in such as manner as to create a staircase upon which they would ascend to unload the cartons from the top down. The in-compliance shipping cartons were never designed to withstand the impact and weight of an adult using them as stairs! Upon witnessing what was happening, the shocked distribution center manager refunded all the supplier's related chargebacks without question.

Several problems are exposed in this story:

1. The requirement of floor-loading seems to be a contributing factor to vendor compliance chargebacks. In general the cause and effect

of requirements should be examined end-to-end with participation by all those involved to root out any unintended consequences.

2. The supplier was forced to incur costs and expend time and resources to investigate and resolve this issue, which was not its fault. The supplier suffered the inability to make use of cash that was incorrectly withheld from it by one of its customers. This could have resulted in delayed projects or postponed hiring. The time spent by the supplier's employees on this matter could have been allocated to other projects. The chargebacks and additional overhead (e.g. travel costs) would have a negative impact on the supplier's financial statements and could thus have had further negative consequences in terms of decisions and assessments based on financial performance.

3. I am left to wonder: How many other suppliers were financially penalized for the same issue but did not have their chargebacks rescinded because they did not make an on-site visit? Did the retailer continue the disparity in assessing chargebacks for this issue or were all chargebacks to all suppliers for crushed cartons as a result of the retailer's own requirements and actions refunded and halted?

The uneven assignment of chargebacks does rather smack of being unethical if not also illegal. It seems considerably unfair that some suppliers get preferential treatment over others. What about suppliers who cannot afford to send their vendor compliance staff – which may be just one person – on a field trip to discover the reasons behind their financial penalties from one retailer to the next and perhaps also from one retailer's distribution center to the next? Are these suppliers just out of luck in terms of being able to refute financial penalties and have money refunded?

According to Section 2-513 (Buyers Right to Inspection of Goods) of the UCC: "Unless otherwise agreed and subject to subsection (3), where goods are tendered or delivered or identified to the contract for sale, the buyer has a right before payment or acceptance to inspect them at any reasonable place and time and in any reasonable manner. When the seller is required or authorized to send the goods to the buyer, the inspection may be after their arrival." Further, the section states that "Expenses of inspection must be borne by the buyer but may be recovered from the seller if the goods do not conform and are rejected."

This removes any question that the retailer – the buying party – has the right to inspect the goods for damage and – as seems to be allowed – conformity to vendor compliance guidelines. However, I would argue that there is a burden on the part of the buyer to be able to accurately and uniformly perform this inspection and levy the penalties fairly. Fraud is defined in this book as a breach of confidence and the supplier must have some confidence that the retailer is being fair in the examination and determination of the shipment, goods, and data being sent for conformity to the established vendor compliance guidelines.

One of the issues at the crux of the chargeback dispute is whether there was in fact a compliance violation at all. Section 2-515 (Preserving Evidence of Goods in Dispute) of the UCC provides clarity to this issue by stating that:

In furtherance of the adjustment of any claim or dispute:

(a) either party on reasonable notification to the other and for the purpose of ascertaining the facts and preserving evidence has the right to inspect, test and sample the goods including such of them as may be in the possession or control of the other; and

(b) the parties may agree to a third party inspection or survey to determine the conformity or condition of the goods and may agree that the findings shall be binding upon them in any subsequent litigation or adjustment.

Because of the rapid pace of the movement of goods to the retail floor, suppliers really have no chance of inspecting their goods at the stores or distribution centers. This would require that retailers retain cartons and have photographic evidence of the original state of the merchandise when it was received from the inbound shipment. Any discrepancy in the goods would require the items to be held in a secure quarantine area until the supplier could perform its own inspection. The extra space and handling considerations alone would likely cost retailers millions of dollars per year and remove them from the sales floor or shelf, possibly causing product damage or expiration, which would be impractical from a performance standpoint though apparently necessary from a legal compliance perspective. But, without evidence, how can a supplier be sure that an assessed compliance violation penalty is correct? Further, how can a supplier be assured that a compliance violation even existed? Perhaps a fictitious violation was assessed to ensure that a quota of compliance chargebacks was reached.

Section 2-607 (Effect of Acceptance; Notice of Breach; Burden of Establishing Breach after Acceptance; Notice of Claim or Litigation to Person Answerable Over) of the UCC seems to provide guidance on the issue of validity in the assessment of chargebacks. What this section covers is that the acceptance of goods does not mean that a remedy for non-conformity of the goods cannot be established, but notes that the "buyer must within a reasonable time after he discovers or should have discovered any breach notify the seller of breach or be barred from any remedy." The delays in chargeback notification to the supplier – which can be weeks or months – and sometimes the convoluted information provided to the supplier are at the forefront of supplier complaints about the chargeback process.

According to the UCC, the question of what constitutes a "reasonable" timeframe is not established, though it may be inferred due to the fact that there are references to conflicts or disagreements being "seasonably cured" or something happening after a "seasonable" action such as a receipt. Again referring to www.dictionary.com, I looked up the word "seasonable" and found two definitions, one referencing "suitable to or characteristic of a season" and the other being something that is either "timely" or "opportune." Without further clarification in the UCC, one could argue that retailers only need to inform suppliers of non-conformity problems at some point within the current season, e.g. winter, spring, summer, or fall. I cannot state that the authors of the UCC meant that the notification was to be timely without regard to the actual season of the year, and the UCC does not further define what amount of time is or is not considered to be "timely."

In July 2011 I received a telephone call from someone in the accounting function at an apparel company in California who sold goods to several very large, well-established US retailers. I was informed that the company had received a letter from one of its retailer customers demanding payment of $57,000 for chargeback violations from two years ago that the retailer's auditors had discovered. The apparel company was looking for help in handling this dispute and responding to the retailer.

Often retailers will specify in vendor contracts that the retailer has the right to automatically deduct penalties for damages of a few percentage points off an order. Defective merchandise beyond that which is damaged is also covered in supplier contracts, as are shared marketing expenses. The supplier – if it has read the contracts before signing – is advised before entering into the supplier–customer relationship of the fees and penalties it can incur.

At the crux of this particular matter is that two years had passed since supposed compliance violations had occurred. I was informed by the apparel company that these fees were not associated with defective merchandise or shared marketing expenses; they were associated with damaged product, which would suggest that the product could not have been sold in the first place.

In discussing this matter with the apparel company, I was told that the sales to this particular retailer had fallen off and they had always been a difficult customer, representing minimal profits and maximum headaches. The apparel company – like many in that industry – factored their invoices, meaning that a third party purchased their receivables; this might have alleviated the apparel company from any financial responsibility from chargebacks depending on the contract with its factor.

After two years, how is it reasonably possible to prove the merchandise was not in fact sold in the retailer's stores? What proof is there that the apparel was damaged? The fact that the retailer would send such a letter believing that it could extract payment from a supplier requires a considerable amount of hubris on its part. I have to wonder what was running through the mind of the executive at the retailer who thought he or she could extract payment from suppliers under such circumstances. Perhaps a vendor compliance violation really did occur – two years ago – but after so long it is my opinion that the retailer must live with the mistake and understand that it missed opportunities to collect compliance violation monies. What is unfortunate is that some current suppliers may actually succumb and pay the purported chargeback amounts out of fear of losing a large customer, akin to an extortion racket.

What's the Verdict?

Taking the law at its face value, the issue of the legality of chargebacks does not appear to be in question. The amount the retailer can rightfully deduct for non-conformity penalties is at the core of the debate: How much does it truly cost to correct some of the non-conformity problems? There do not appear to be any retail industry standards on this issue which would help to establish some uniform ground rules. If one retailer's vendor compliance program runs less efficiently than that of another retailer and thus charges more for non-compliance penalties to compensate for an inefficient operation, is this fair to the suppliers? This would in all likelihood be a very good project for the retail industry as a whole to tackle and establish standard deduction amounts in concert with standard terminology and deduction reasons, both of which can vary from one retailer to the next.

The validity of chargebacks in terms of their amounts and assessment seems to be the greater problem and one that is more open to fraudulent behavior. Retailers must accept the responsibility of being able to accurately monitor supplier behavior and correctly determine and communicate non-conformity issues to the suppliers. If a retailer cannot assess vendor compliance chargebacks with the absolute assurance of being correct, then in my opinion the retailer needs to rethink its chargeback policy and withdraw from withholding supplier money for anything but the most obvious violations, e.g. a missing carton label. Ethically, no supplier should be able to be excluded from the penalties of non-compliance if in fact that supplier is guilty of not conforming to the requirements that were fully disclosed upon consideration of the business relationship and prior to the formalization of the relationship. The competitive nature of retail and consumer packaged goods should not be able to breach the basics of good ethical behavior.

One considerable sticking point is the supplier's right to inspect goods that the retailer believes are "damaged" and do not conform to the stated and agreed-upon vendor compliance requirements. (It should be kept in mind that the issue of "damages" extends to all aspects of vendor compliance and includes quality problems related to the items, data accuracy and integrity of EDI transactions, barcode labels on cartons and pallets, properly formatted packing lists and bills of lading, properly stacked pallets, the use of proper or designated hangers, properly formatted item hangtags with accurate price and description information, and more.) While how much a retailer should charge as a penalty for non-compliance is a matter of debate, the right of the supplier to inspect goods deemed damaged is a matter of fact, though it is a right that the supplier must be aware of and invoke according to the letter of the law. However, if the supplier is notified too long after the goods have been received and distributed through a retailer's supply chain – and perhaps even sold – the supplier is left without the ability, even upon notification, to inspect the goods to confirm or deny conformity to the retailer's vendor compliance guidelines. Beyond everything else related to this discussion, it is on this issue that I believe retailers are not in compliance with the letter if not also the spirit of the UCC.

Most organizations have mission-type statements that promote honesty, integrity, and ethical behavior in how they operate and in their customer, employee, supplier, and competitor relationships. These corporate statements should reflect the control environment (the "tone at the top") of the organization and should act as the mantra for how the organization acts at all levels. If retailers are on the whole acting true to what they profess in their mission statements, then what is the source of the sour relationship between suppliers and retailers with regard to vendor compliance? If the retailers are promising to follow the law and act ethically, how come it appears that they are not and why don't they just do something about it and fix the problems?

"Authority leakage" is a term used to describe the distance between a level of management and the knowledge of the actual business operations. It is a disconnect between what is perceived to be happening and what is really going on. At the top levels, organizational management may believe they are operating with full integrity and upholding the highest levels of ethical behavior, while in reality this is not the case. Good governance practices require management to understand what their subordinates are doing, and if the right information is being communicated up through the hierarchy, executive management would be aware of business practices that might be considered questionable in nature if they are not obviously flawed.

The distinction is important when considering whether illicit behavior is occupational (perpetrated by one or more persons using their job roles to their advantage) or organizational (perpetrated by executives in their job roles of commanding the organization to perform based on their direction) in nature. Fraud is carried out with intent, such as deliberately not providing or masking critical information to supervisors, especially when the information is important to the determination of lawful or illegal activity. In circumstances where the information – including that related to policies or procedures – is either not passed along to executives or is masked so as not to reveal its true nature, fraudulent behavior may be covered up. If everything was being done "by the book," there would be no need to disseminate partial truths or manipulated math, unless the tone at the top is such that an employee fears that the truth will result in job loss or if such behavior is actually rewarded. Executive management at retail companies may truly believe that their organizations are acting in a lawful and ethical manner – and I will give them the benefit of the doubt – but in reality, from my experience with my former employers and clients, this is simply not the case.

This takes us back to the accusation by the general supplier community that retailers are using vendor compliance chargebacks as a profit center, assigning penalties when there are none and/or assessing amounts in excess of the cost of the correction. Retailers do make errors in judgment and may assign chargebacks too quickly without investigation, as the example set by the story of the shoe manufacturer that witnessed its shipping cartons used as stepping stones illustrates. Withholding money under what would be false pretenses would fit the consideration of fraud, and retailers that do this until the supplier complains and personally investigates are guilty of perpetrating illicit and unethical behavior.

While not all retailers are guilty of other-than-good behavior, there is plenty of blame to be assigned to a significant percentage if not the majority of retailers. Would either the RICO and/or the Sherman Anti-Trust Act be in violation by the retail industry even if retailers did not officially convene to conspire against suppliers by deliberately not coming together the improve the vendor compliance process? Does deliberately convoluting a process that withholds money due to another party and acts to obscure its operational details translate into fraudulent behavior? These questions are better left for those in the legal profession. The point is that there is generally an acrimonious relationship between the retailer and supplier communities. This is due to the somewhat secretive operations of retailer–vendor compliance programs and a reluctance on the whole by retailers to come together to address the problems. Fraud is carried out with intent, and intending to not address a problematic process that results in financial penalties could be considered unethical as well as fraudulent behavior. With a law – the UCC – that lays out the foundation for the buyer–seller relationship, it would seem to make sense to base vendor compliance programs on the letter of the law and the spirit of good ethical behavior, as promised in some retailer mission statements.

During a consulting assignment where I was helping a retail supplier minimize chargebacks and improve operational performance, I happened to be at the supplier's office when we received a certificate in the mail from one of my client's retail customers that graded supplier performance every six months. The certificate, scorecard, and accompanying letter were representative of achievement in being a top-performing supplier, which came with a particular perk: Even if my client did something that would cause a chargeback, there would be no financial penalty through the current scorecard period because – as the retailer stated in the letter – we are all human and everyone makes honest mistakes now and then. Wow.

During the course of the current grading period, my client did make a few mistakes and, true to its word, the retailer did not assess financial penalties because of my client's current top-performer status. It is difficult to argue against a vendor compliance program that recognizes human error and rewards suppliers that are willing to take steps to do better.

When there is bad behavior perpetuated by a majority of organizations – either by number or sheer size as a representation of the overall market share – consumers are likely to be left to suffer at the end via higher prices to cover excessive customer and supplier operating costs. Because the practice is so widespread, it can be considered an *industrial* problem even if not all the players are accomplices. Vendor compliance costs individual suppliers from thousands to millions of dollars per year in excessive expenses that are eventually passed along in the price as they become part of the cost of goods sold when factored into the supplier's operating costs. The retail industry's lack of uniform standards versus guidelines is a contributing factor toward the complexity of compliance. At the very least, uniformity in compliance chargeback reasons and amounts would be a great start to reducing the confusion and would help to inject some confidence into the supplier–retailer relationship.

There is little question that vendor compliance chargebacks are legal: The UCC seems clear in permitting them. And I would agree that the UCC allows for compensation beyond the cost of the correction, e.g. administrative and monitoring costs. However, I do not agree that the financial penalties being assessed are an accurate reflection of the cost of the corrective actions. I certainly do not agree with the basic premise of vendor compliance programs that the vendor is guilty until proven innocent, especially when the ability of the supplier to prove itself – as allowed by the UCC – is rendered futile by the actions of the retailer. As the sole decider of whether a supplier is financially penalized, the retailer must apply the law to all suppliers uniformly without distinction and must make every effort to ensure the accuracy of its decision-support systems and the people involved. Only with integrity in the process can there be fairness in the application of the punishment.

The Importance of Vendor Compliance Programs in Fraud Detection

Operated legally and ethically, vendor compliance programs are an effective means of bringing operational efficiencies to the table as well as providing a foundation for the detection and determination of suspected fraudulent activities by suppliers and the employees and contractors of customers, in collaboration and acting on their own. Remember: The supply chain is internal and external, and the same (or at least nearly similar) criteria for external supplier performance can be applied to internal suppliers (e.g. departments) within an organization.

From the perspective of good governance, vendor compliance programs should exhibit the following characteristics:

1. They should be constructed as such to follow all applicable laws such as the UCC in the US.

2. They should apply equally to all suppliers without exception.

3. The customer should implement the necessary computer applications, processes, and education programs to enable its personnel – at all levels of the organization – to be able to fairly and uniformly monitor and judge a non-conformity violation. Employees should not be incentivized to find fault; the customer organization must find another way to judge the performance of its employees involved in administering its vendor compliance program.

4. Penalties for compliance violations should reasonably cover the cost of the correction and administrative actions and should not be excessive such that the customer is unfairly profiting from it vendor compliance program.

5. It should be remembered that we are all human and that we all can make honest mistakes. If the customer cannot accurately make a determination about non-conformity, it should not. Suppliers should be granted reasonable leeway for occasional and unintentional mistakes.

It is very important to recognize that good supply chain relationships are about partnerships, whether internal or external to the organization. Supply chain partners help one another through the good times and the bad, and can be critical to success in times of crisis. The "trust but verify" mentality helps keep both the supplier and customer entities aware that the other is watching the relationship, albeit from its own perspective. As I previously stated, organizations need to ensure that their customers, suppliers, and employees are able to report suspected fraudulent activities anonymously. Claims based on fact should require shorter investigation timeframes to resolve.

The vendor scorecard – based on metrics against key performance indicators – can be used as a foundation for the determination of fraudulent behavior as it sets a baseline for what should be (achievable) normal behavior. If a customer has a disruptive supplier, is it fraud or just laziness? A supplier who fulfills orders continually short may just have poorly run operations or may have a rogue employee inside and be either unaware or perpetrating organizational fraud. Do the shipping cartons look like they are being opened and resealed? If so, then perhaps the freight carrier is to blame. Ruling out those possibilities, the investigative trail may lead to the customer's own operations where perhaps there are illicit activities on the receiving dock or in inventory control. Conversely, a supplier (internal or external) that always performs at a 100 percent level may truly be great at what it does or there may be collusion and a cover-up of the supplier's shoddy performance or intentional deception.

In establishing metrics (the measurements against which performance is judged) and the key performance indicators (the level of performance against a metric), the vendor scorecard should include the following behavioral assessments:

- Order acceptance: The number of and frequency of cancelled purchase orders or work orders depending upon which document is relevant to the internal or external supplier.
- Fulfillment accuracy: The measure as to whether all the goods demanded were delivered in their full quantity, on time, and in good condition.
- Invoice performance: Was the invoice submitted by the external supplier done so accurately and on a timely basis?
- Receiving accuracy: Whether the buyer/recipient of the goods received them accurately and, if necessary, whether the receipt was reported to the supplier on a timely basis.
- Returns/replacements: An analysis of quality as reflected by product returns or replacements.

The fact is that in order to determine fraudulent activity, we need a baseline for good behavior. Sometimes these baselines are in the form of laws that tell us what is or is not acceptable behavior. Other times we might look toward societal norms to guide us. The application of the Golden Rule – "Do unto others as you would have them do unto you" – helps to define behavior based on how we ourselves would like to be treated or not to be treated. Vendor compliance programs that are fair and reasonable are generally acceptable by suppliers that recognize the need for them and offer opportunities for self-improvement. The behavioral baseline must be a combination of acting ethically and assessing accurately. It is when the rules of the game are abused or selectively enforced that relationships between partners become strained to the point of animosity, making reparations difficult as each party holds to its own perspective more fiercely as the relationship deteriorates further. People and organizations that adhere to good governance must "do the right thing" and that sometimes means doing something other than what you would prefer.

The April 17, 2011 edition of the *South Florida Sun Sentinel* contained a half-page advertisement by Baptist Health South Florida, which stated that it was the "largest faith-based not-for-profit health care organization" in the geographic area that includes the major cities of Miami and Fort Lauderdale as well as the Florida Keys. Baptist Health was providing notification of being recognized as one of the 2011 "World's Most Ethical Companies" by the Ethisphere Institute (www.ethisphere.com).

Baptist Health's advertisement contained two statements in large font: The first statement proclaimed "It's not only what you do." The second statement proclaimed "It's how you do it." Those two statements together are a fine representation of what ethical behavior is all about: The process matters as much as, if not more so than, the results. How you get to where you are going is as important, if not more important, than the destination itself.

Checking the Ethisphere website, I saw an interesting graph titled "Percent Returns – World's Most Ethical (WME) Companies vs. S&P 500." Ethisphere states "It can pay to be ethical" and continues that "Investing in ethics is beneficial for any company, even in a recession." According to Ethisphere: "WME companies, if indexed together, have routinely and significantly outpaced the S&P 500 each year since the recognition's 2007 inception. On average, the WME companies outperformed the S&P 500 by 7.3 percent annually. In addition to increased financial performance, ethical companies benefit from better brand reputation, consumer loyalty and higher employee retention rates."

The end goal of vendor compliance can be thought of as a protection of the Customer's Bill of Rights mentioned earlier in this book. The customer's rights can be trampled upon by operational disruptions, whether due to inefficiencies or fraudulent behavior. Protection of the customer's rights – no matter whether the customer is internal or external (e.g. the consumer) – is a holistic perspective that should be embraced at every level of an organization. If at each step along a supply chain the supplier exhibited the highest level of care and concern that it passed on perfect product (raw materials, components, finished goods, documents, information) to its customer, I would argue that the result would be an organization that operated with great integrity and great efficiency, and that the organization would employ some of the most satisfied people around.

The more engaged the employee and the more engaged employees that an organization retains, the healthier the organization will be. And a healthy organization is one that can be more effective at keeping fraud at bay.

PART V

Fraud Detection: Past, Present, Future

21

Predicting the Future of Supply Chain Fraud

My earliest memories growing up were of reading the daily newspaper. I don't know why, and I don't think I was forced to by my parents, I think it's just something I did. Even in my teenage years, if a day went by and I did not get my dose of daily news, I felt strangely disconnected. Through my writing and speaking, people often ask me how I know what I know about what goes on in the world, and my answer is that I simply read my daily newspaper and approximately one dozen publications I subscribe to covering subjects such as finance, information, technology, fraud, consumer products, retail, and supply chains.

As you have probably noticed, many of the analogies used in this book originate from South Florida, comprising the areas around the major cities of Miami, Fort Lauderdale, and West Palm Beach, where I have lived since 1992. Florida, the fourth most-populous state in the US, and in particular the South Florida area, consistently rank at the top of several "worst" lists, such as being in the top ten geographic areas for mortgage fraud and ranking at or very near the top in Medicare billing fraud. From 2007 to 2009 Florida ranked first in the US for insurance fraud complaints related to staged (fictitious) automobile accidents. Insurance companies reported just over 3,000 suspected fraudulent claims in Florida, a little under the number of suspected claims in California and New York – the next two states in the ranking – combined, with the South Florida area leading the state. In 2011 the FBI busted the largest Medicare billing fraud scheme to date involving over 100 people responsible for over $225 million in fraudulent billings. The FBI has also arrested dozens of public sector officials – many in South Florida – for what can be summarized as public corruption, such as taking bribes or kickbacks to vote in favor of awarding contracts to certain companies and failing to disclose close – sometimes even family – relationships during the bid and awarding processes. There is just no shortage of fraud schemes being hatched in my own backyard. As a fraud fighter, I should be very happy that I do not have to look far for examples, but I am dismayed at the sheer number and size of the fraud schemes that I read about.

In May 2010 I was contacted by a meeting and conference company in Mumbai, India who found my www.supplychainfraud.com website and was so impressed with the business model that they wanted to make it into a two-day conference with me as the keynote speaker, as well as providing other presentations. Being the analytical, cautious, and somewhat suspicious person I am, I thought it was a joke or an attempt to scam me for money. I also thought that if this was an attempt to get me to fly literally halfway around the world to kidnap me and hold me to ransom, my kidnappers would be sorely disappointed at how little money they would get for my return. Nonetheless, I have learned that opportunities not explored are opportunities that are missed and so thought that there was no harm in sending back a reply email stating my initial interest. We arranged to chat thanks to voice over the Internet and they put the conference together over the next few months, inviting an impressive line-up of local (Mumbai-based) speakers discussing topics including fraud in warehouse operations, protecting technology assets, and fraud investigation techniques.

I admit that as the conference was approaching, I was wondering how my sense of humor – spontaneous, observational, sarcastic without being truly mean-spirited – would be accepted by the attendees. My sense of humor generally plays out very well to both American and British audiences and I regularly use humor during presentations to keep the audience's attention. (I have been actively engaged in the British-American Chamber of Commerce in South Florida since 2005 and have had many opportunities to spend time with ex-patriots from the UK.) It's not that I feel that my material itself cannot hold an audience's interest, but I make sure that the delivery maintains its professionalism without becoming boring.

My keynote presentation on detecting and reducing supply chain fraud was the first one on the morning of the first day. (My presentations on good governance for supply chain operations and creating effective vendor compliance guidelines were featured on the second day of the conference.) I delivered my presentation just as I had numerous times before and the audience sat quite stone-faced even through the humorous bits. Granted, morning presentations can be tough if the audience isn't awake yet, but I felt that something else was amiss. Not too far into my presentation, I was thinking that it was going to be a long conference. I finished my presentation and opened the floor to questions. One of the first questions to be asked was prefaced by the attendee with what almost amounted to an apology. The gentleman effectively stated that business practices in India were ripe with fraud, known for fraud, that India was behind the times in addressing fraud, and it was of little wonder that India was not viewed as a peer among other global leaders such as the US despite its status as an emerging economy.

As I listened, I quickly realized that it was very likely that the audience perceived me to be an arrogant American who came halfway around the world to tell the attendees how things are (supposedly) done – and to perfection – in the US, where, in the minds of my Indian audience, fraud had been all but completely eradicated, and how less-advanced nations like India need to do better to be taken seriously as true business partners.

Having a revelation and seizing the moment, I prefaced my answer by first leveling the playing field: I literally told the audience that, likely contrary to their perception, the US had not only basically perfected fraud – probably inventing many of those plaguing the world today – but did it on a bigger and more extensive scale than anyone else in the world. I continued by saying that the monetary impact of frauds in the US is measured in billions, not merely millions or thousands, and that emerging countries like India were merely rank amateurs when compared to the professional fraudsters we have in the US. The audience burst out laughing and I knew instantly that we had connected with each other.

Throughout the rest of the conference, I enjoyed chatting with the attendees and other speakers on personal and professional matters, joking back and forth as freely as I would with people I have known for a much longer period of time. The attendees knew that I was there to share my experiences and information to help them do better and not to tell them how good or bad a job had been done up to that point in fighting fraud in India.

The conference was attended by over 50 delegates from public companies, private corporations who sent various managers, directors, chief security officers, and vice presidents, as well as various Indian government agencies that sent their chief vigilance officers (CVOs). From what I learned, the CVO role is relatively new, the position created at government agencies as mandated by laws recently passed. Acting as a somewhat hybrid mix of internal audit and internal affairs, the CVO's office provides an outlet for the reporting of suspected illicit activities perpetrated by agency employees and suppliers. The CVO office performs due diligence to ensure the agency's operations conform to what those of us in the US would consider to be good governance practices. Each Indian government agency – and, from what I could establish, each local office of every Indian government agency – has a CVO.

The functions of the CVO include the detection of fraud, the preparation of documents investigating the allegations, and assisting in the punishment of the guilty. But the purpose of the CVO is not intended to be merely reactive but also – and more so – proactive in the establishment of preventative measures such as unannounced inspections and the identification and monitoring of areas known to be fraud hotspots. Key areas of focus for the CVO include gifts provided to and accepted by public officials, the employment of relatives of a government agency employee by a private-sector firm, and, I strongly suspect, whether the firm has the potential of or is currently conducting business with a government agency.

In July 2011 auditors from the state of Florida revealed the discovery that over a two-year period, from 2008 to 2010, 574 contracts valued at over $55 million went to companies either controlled by or linked to board members from the 24 regional offices of the state's labor development network, which is designed to provide training and job assistance to unemployed Floridians. (The US Department of Labor has also launched a probe into the practices of the state workforce boards because Florida receives approximately $250 million per year from the federal government.) While the bulk of those funds went to non-profit organizations, over $7.7 million went to private companies. Each of the 24 regional office boards are composed of 30 or more members, including heads of local universities, various government agencies, and local businesses. According to the article in the *South Florida Sun Sentinel* newspaper: "State law at the time required that contracts with board members be approved by two-thirds of the agency's full board, but that often didn't happen. In a few instances, auditors were unable to verify that board members who profited from a deal disclosed a conflict or abstained from voting."

The supporters of this conduct quoted in the article point toward cases where the lowest bidder won the contract and it just so happened that the lowest bidder was or was linked to a board member. One defender of the practice stated: "If you're the low bidder, should you be excluded?" But one overriding question I have is whether the goods and services provided were even necessary, or was the fact that a board member represented a company that could provide a product or services enough to decide who should be awarded the business? Unless the goods or services to be procured are absolutely necessary to the functioning of the agency, unless full disclosure as to personal or professional relationships are made, unless relative distanced is exercised whereby the person involved excuses himself or herself from the vote, and unless the other board members can cast an honest vote without regard to their colleague's situation and financial gain, and can do so without fear of repercussions should they be the person involved the next time, the answer is quite simply "Yes, board members should be excluded." This is because, aside from the violations of good governance, the primary purpose of being on a board of directors of a government agency should not be to secure business or oneself or one's colleagues. While the investigation seems to center on contracts to private companies, I believe that the contracts to the non-profit organizations bear scrutiny too, because non-profit organizations can still be the recipient of public funds. My question as to whether the products and services purchased were even necessary does not differentiate between where they were purchased from (non-profit organization or private company) so long as there was a link – direct or indirect – to a workforce agency board member. I think that the state of Florida could learn a thing or two from India in establishing a CVO's office wherever public funds are being spent.

Also in July 2011 the US government launched a nationwide computer system that uses predictive modeling to detect patterns and anomalies that could represent fraud. The system was originated and tested in the South Florida area, which represented $1.85 billion out of $2.3 billion in fraudulent Medicare billing schemes exposed by Medicare fraud strike forces – netting over 1,000 defendants – going back to March 2007. This new computer system can help root out criminal networks and complicit patients that operate all over the country and perpetrate Medicare and medical billing frauds, and well as identify potential identity theft and analyze health care providers to exclude any suspected of fraud. It is estimated that as much as $60 billion per year on the whole – one-tenth of the total annual spending for Medicare – is lost to fraud.

While Medicare claims have been reviewed in the past for obvious anomalies, such as a man seeking reimbursement for a medical procedure that would be only applicable to a woman, this method was reactive in that the claim was paid and then only possibly caught if reviewed. The new method of claims review is based upon the concept of predictive modeling and "began looking a little deeper, using sophisticated computer analysis of billing pattern" according to Peter Budetti, who leads the anti-fraud initiative for Medicare. Budetti further states: "We can't prosecute our way out of this. We have to prevent it. We have to prevent these bad guys from getting into the program, preventing money from getting out the door."

A white paper by Accenture titled "How Predictive Analytics Helps Supply Chains Compete and Win in Today's Volatile Business Environment" describes two types of analytics: descriptive and predictive. Descriptive analytics, according to the white paper, looks at historical data and, using queries, reports, and alerts, "helps companies answer basic questions such as what happened, why did it happen, and how much did it hurt us or benefit us?"

Predictive analytics, the white paper states, "combines historical descriptive analytics with more sophisticated statistical modeling, forecasting, optimization, etc., to anticipate the impact on business outcome. Predictive analytics helps companies answer questions like why is this happening, and what's the best that could happen given the decision support that the analytical analysis has provided?" The white paper highlights assessing supply chain risk as a key capability of predictive analytics because it "enables managers to identify where problems are likely to arise along the supply chain, highlighting critical areas for risk mitigation." Regardless of whether it is perpetrated internally, externally, or a combination of both, and regardless of whether it involves raw materials, components, finished goods, documents, monies, or services, fraud in supply chains represents a serious risk to the well-being of an organization.

The following information is from the Predictive Analytics World website (www.predictiveanalyticsworld.com) and highlights some of behaviors that can be predicted from different business applications (Table 21.1).

Table 21.1 Business Applications and Predicted Behaviors

Business Application	What Is Predicted
Customer Retention	Customer Defection or Attrition
Direct Marketing	Customer Response
Product Recommendations	Customer Likes & Wants
Behavior-based Advertising	Which ad the customer will click on
Email targeting	Which message customer will respond to
Credit scoring	Debtor risk
Fundraising for non-profits	Donation amount
Insurance pricing and selection	Applicant response, insured risk

Before we can get on to predictive analytics, the descriptive analytics must be achieved. Descriptive analytics must rely on collected data with the more data collected, the better. Gaps in data collection can render descriptive analytics incomplete, meaning that something (e.g. particular transactions such as purchases, payments, or inventory movements) cannot be analyzed because the data does not exist to do so. Likewise, if the data does exist but is not delivered to the business analytics software on a timely basis, it may not be included in the analysis. This data integrity – mentioned in the review of the COSO compliance framework – is very important because anything less would reflect inaccurate information and could lead to false conclusions about the whats, whys, and hows of historical events. Just as descriptive analysis looks into the past, we can look at existing long-used technologies to close data gaps. Ensuring that all the necessary data is collected returns us to the use of automatic identification (e.g. barcode scanning), EDI, and other forms of eB2B that integrate with the ERP system or other business software backbone to convert paper-based transactions to paperless electronic transactions and thus have the data available for the analytic software. The integrity of the data is validated through audits and cross-checks against related transactions and data tables within the secure systems and applications environment as reviewed in the discussion of the COBIT compliance framework.

In the July 11, 2011 edition of *Information Week* magazine there is an article entitled "The Data Mastery Imperative" written by Michael Sharpe. The article begins: "There's a disconnect between the emphasis companies place on real-time metrics and the amount of time IT organizations spend developing policies and procedures for management the data that comprises those metrics. Everyone wants to turn raw stats into wisdom in real time, but you can't even begin to do that if you don't have a handle on your data." Mr Sharpe stresses that without business processes – including data collection and governance – in place to maintain consistent data, embarking on a master data management (MDM) project is pointless.

What I found interesting in the article is that Mr Sharpe discusses the importance of establishing the producers (who I have termed *suppliers*) and consumers (who I have termed *customers*) of data, as well as defining who owns and holds accountability for different sets of data. As data such as a customer record follows a supply chain path, it may change, and this change must be controlled and governed accordingly because the results could taint analytical output. For example, a customer data record containing name and address information sits in the customer master file. The customer orders goods or services from a company and, during the sales order entry process, the customer either informs the sales representative – or makes the change himself or herself, such as on a web portal – of a new address. A customer with one address in his or her master data record and another address on a sales order is just one of the disconnects that can happen without a cohesive master data management plan in place that embodies both technology and business processes.

A change in a customer address may be somewhat innocuous, but what if prices were changed in favor of the customer to perpetrate a fraud scheme? This could be accomplished via collusion with a sales representative or in concert with someone in the technology department who could, for example, insert program code that permitted price changes by unlocking a data field on a web form for specific customer account identifiers. Without outside-the-application audit programs, this fraud would likely go unnoticed for a long time.

Failing to protect the integrity of data as it moves along the supply chain path can result in a failure to identify an anomaly that is occurring. For US public companies that must adhere to SOX, this can result in inaccuracies that find their way into company financial statements. Simply reporting financial statements on a timely basis is not enough; they must be reported.

Parameters for performance will tell us if the numbers produced from the descriptive analysis indicate a problem or not. The establishment of metrics requires the definition of characteristics of business performance that are important in determining the health of the organization. The number of new customers acquired, the number of products returned, the accuracy of orders shipped, the level of inventory held – these are all metrics that characterize a business, and how they are calculated and from which sources of data must be defined. Setting key performance indicators against the metrics turns the metrics data into meaningful information, e.g. 2.5 percent of a particular category of products were returned in the last quarter. Is that good or bad? Establishing a scorecard rating criteria makes an objective number more subjective to the business – there might be a high rate of return in one industry but a low rate of return in another industry.

While all that helps to keep the bad data out, the use of descriptive analytics can help us keep the bad "tangible stuff" out of our supply chains. If the performance of a supplier – whether internal or external – should fall below a minimum acceptable standard (the supplier's scorecard rating drops), then the quality review of the supplier's deliverable (whatever that may be) should increase accordingly. We may not be able to predict that there will be any problems, but we can use history from descriptive analytics as a guide to quickly – if the right information is available at the right time – adjust to a changing condition and take extra preventative measures. An improvement in supplier performance will be a good sign, but a higher than normal level of verification should remain until trust is fully restored to the relationship.

Can fraud be predicted? This is analogous to asking if human behavior can be predicted. If we know how – in a given situation – a person is likely to respond, then the answer is "yes," but only sometimes. In the detection and reduction of supply chain fraud, we can use descriptive analytics to help us find the gaps where fraud can exist and highlight abnormal behavior that warrants further investigation to us. Organizations cannot rest on their laurels or be complacent because they have not been – or believe they have not been – the subject of fraud. And once counter-measures have been put into place, they need to be continually monitored and adjusted as the business of the organization changes and new technologies and methodologies become available. Fraud detection and reduction is not a one-time event, it is something to be continually monitored to verify that it is adequate for the organization.

We cannot become our own worst enemy and allow pride or politics, greed or graft to get in the way of doing the right thing, whether it is behavioral, procedural, or systematic. Fraudsters will exploit weaknesses and take advantage of situations until we each achieve the right tone, connect the dots, close the gaps, and shut them out. Organizations that do not address the risk of fraud or operate in an unethical manner may one day find themselves to be the subject of a self-fulfilling prophecy as the victims of frauds that they failed to discover or allowed to be perpetrated. That is what I predict.

Online Resources

60 Minutes
Available at: http://www.60minutes.com [accessed: 31 March 2012]

Accenture
Available at: http://www.accenture.com [accessed: 31 March 2012]

Agriculture of the Middle
Available at: http://www.agofthemiddle.org [accessed: 31 March 2012]

American Numismatic Association
Available at: http://www.money.org [accessed: 31 March 2012]

Aon
Available at: http://www.aon.com [accessed: 31 March 2012]

Associated Press
Available at: http://www.ap.org [accessed: 31 March 2012]

Association of American Colleges and Universities
Available at: http://www.aacu.org [accessed: 31 March 2012]

Broward Bulldog
Available at: http://www.browardbulldog.org [accessed: 2031 March 2012]

Canisius College
Available at: http://www.canisius.edu [accessed: 31 March 2012]

Center for Integrated Agricultural Studies – University of Wisconsin (Madison)
Available at: http://www.cias.wisc.edu [accessed: 31 March 2012]

CFO.com
Available at: http://www.cfo.com [accessed: 31 March 2012]

China Internet Information Center
Available at: http://www.china.org.cn [accessed: 31 March 2012]

Committee of Sponsoring Organizations of the Treadway Commission
Available at: http://www.coso.org [accessed: 31 March 2012]

Cornell University Law School, Legal Information Institute
Available at: www.law.cornell.edu [accessed: 31 March 2012]

CSO Online
Available at: http://www.csoonline.com [accessed: 31 March 2012]

Cyber-Ark
Available at: http://www.cyber-ark.com [accessed: 31 March 2012]

Ethisphere
Available at: http://www.ethisphere.com [accessed: 31 March 2012]

Fairleigh Dickenson University
Available at: http://www.fdu.edu [accessed: 31 March 2012]

Food Logistics
Available at: http://www.foodlogistics.com [accessed: 31 March 2012]

Gallup Management Journal
Available at: http://gmj.gallup.com [accessed: 31 March 2012]

Hoover, J.N., 2009. "Secure the Cyber Supply Chain." *Information Week* 9 November, pp. 45–52.
Available at: http://www.informationweek.com [accessed: 31 March 2012].

Inbound Logistics
Available at: http://www.inboundlogistics.com [accessed: 31 March 2012]

Information Systems Audit and Control
Available at: http://www.isaca.org [accessed: 31 March 2012]

Information Week
Available at: http://www.informationweek.com [accessed: 31 March 2012]

The Institute of Internal Auditors
Available at: http://www.theiia.org [accessed: 31 March 2012]

Internal Revenue Service
Available at: http://www.irs.gov [accessed: 31 March 2012]

International Standards Organization
Available at: http://www.iso.org [accessed: 31 March 2012]

Iowa State University
Available at: http://www.iastate.edu [accessed: 31 March 2012]

Josephson Institute
Available at: http://josephsoninstitute.org [accessed: 31 March 2012]

MSNBC
Available at: http://www.msnbc.com [accessed: 31 March 2012]

The National Archives, Fraud Act 2006
Available at: http://www.legislation.gov.uk/ukpga/2006/35/contents [accessed: 31 March 2012]

National Archives and Records Administration
Available at: http://www.archives.gov [accessed: 31 March 2012]

National Conference of Commissioners on Uniform State Laws, Uniform Law Commission
Available at: http://www.nccusl.org [accessed: 31 March 2012]

National Motor Freight Traffic Association
Available at: http://www.nmfta.org [accessed: 31 March 2012]

Numismatic Guaranty Corporation
Available at: http://www.ngccoin.com [accessed: 31 March 2012]

Predictive Analytics World
Available at: http://www.predictiveanalyticsworld.com [accessed: 31 March 2012]

Reuters
Available at: http://www.reuters.com [accessed: 31 March 2012]

Reverse Logistics magazine
Available at: http://www.rlmagazine.com [accessed: 31 March 2012]

Supply & Demand Chain Executive
Available at: http://www.sdcexec.com [accessed: 31 March 2012]

Time magazine
Available at: http://www.time.com [accessed: 31 March 2012]

Tulane University Law School
Available at: http://www.law.tulane.edu [accessed: 31 March 2012]

United Nations Economic and Social Commission for Asia and the Pacific
Available at: http://www.unescap.org [accessed: 31 March 2012]

United Nations Economic Commission for Europe, *United Nations Directories for Electronic Data Interchange for Administration, Commerce and Transport*
Available at: http://www.unece.org [accessed: 31 March 2012]

United Press International
Available at: http://www.upi.com [accessed: 31 March 2012]

United States Department of Defense
Available at: http://www.dod.gov [accessed: 31 March 2012]

United States Department of Justice – Antitrust Division, 2005. *An Antitrust Primer For Federal Law Enforcement Personnel,* Washington, DC: United States Department of Justice
Available at: http://www.justice.gov/atr/public/guidelines/209114.htm [accessed: 31 March 2012]

United States Department of Justice
Available at: http://www.justice.gov [accessed: 31 March 2012]

United States Department of Labor
Available at: http://www.dol.gov [accessed: 31 March 2012]

United States Federal Bureau of Investigation
Available at: http://www.fbi.gov [accessed: 31 March 2012]

United States Food and Drug Administration
Available at: http://www.fda.gov [accessed: 31 March 2012]

United States Securities and Exchange Commission
Available at: http://www.sec.gov [accessed: 31 March 2012]

United States Sentencing Commission
Available at: http://www.ussc.gov [accessed: 31 March 2012]

United States Supreme Court
Available at: http://www.supremecourt.gov [accessed: 31 March 2012]

Voluntary Interindustry Commerce Solutions Association
Available at: http://www.vics.org [accessed: 31 March 2012]

Wall Street Journal
Available at: http://online.wsj.com [accessed: 31 March 2012]

World Trade 100
Available at: http://www.worldtradewt100.com [accessed: 31 March 2012]

Zetes
Available at: www.zetes.com [accessed: 31 March 2012]

Index

Note: page numbers in *italics* refer to figures; those in **bold** to tables.